NON-DESTRUCTIVE TESTING OF FIBRE-REINFORCED PLASTICS COMPOSITES

VOLUME 1

Edited by

JOHN SUMMERSCALES

*Structural Testing Facility,
Department of Thermofluids and Mechanics,
Royal Naval Engineering College,
Manadon, Plymouth, UK*

MICHIGAN MOLECULAR INSTITUTE
1910 WEST ST. ANDREWS ROAD
MIDLAND, MICHIGAN 48640

ELSEVIER APPLIED SCIENCE
LONDON and NEW YORK

ELSEVIER APPLIED SCIENCE PUBLISHERS LTD
Crown House, Linton Road, Barking, Essex IG11 8JU, England

Sole Distributor in the USA and Canada
ELSEVIER SCIENCE PUBLISHING CO., INC.
52 Vanderbilt Avenue, New York, NY 10017, USA

WITH 18 TABLES AND 95 ILLUSTRATIONS

© ELSEVIER APPLIED SCIENCE PUBLISHERS LTD 1987

British Library Cataloguing in Publication Data
Non-destructive testing of fibre-reinforced
plastics composites.
Vol. 1
1. Fibre reinforced plastics—Testing
2. Non-destructive testing
I. Summerscales, John
668.4'94   TP1177.5.F5

Library of Congress Cataloging in Publication Data
Non-destructive testing of fibre-reinforced plastic
composites.

Includes bibliographies and index.
1. Non-destructive testing.   2. Fiber reinforced
plastics—Testing.   I. Summerscales, John.   II. Title:
Fibre-reinforced plastic composites.
TA417.2.N6565   1987   620.1'127   87-5274

ISBN 1-85166-093-3

No responsibility is assumed by the Publisher for any injury and/or damage to persons or property as a matter of products liability, negligence or otherwise, or from any use or operation of any methods, products, instructions or ideas contained in the material herein.

Special regulations for readers in the USA
This publication has been registered with the Copyright Clearance Center Inc. (CCC), Salem, Massachusetts. Information can be obtained from the CCC about conditions under which photocopies of parts of this publication may be made in the USA. All other copyright questions, including photocopying outside the USA, should be referred to the publisher.

All rights reserved. No part of this publication may be reproduced, stored in a retrieval system, or transmitted in any form or by any means, electronic, mechanical, photocopying, recording, or otherwise, without the prior written permission of the publisher.

Printed in Great Britain by the University Press, Cambridge

NON-DESTRUCTIVE TESTING OF FIBRE-REINFORCED PLASTICS COMPOSITES

Volume 1

Contents

List of Contributors vii

Introduction ix

1. Radiography 1
   A. F. BLOM and P. A. GRADIN

2. Acoustic Emission 25
   M. ARRINGTON

3. Thermal NDT Methods 65
   K. E. PUTTICK

4. Optical Methods 105
   COLIN A. WALKER and JAMES MCKELVIE

5. Vibration Techniques 151
   P. CAWLEY and R. D. ADAMS

6. Corona Discharge 201
   JOHN SUMMERSCALES

7. Chemical Spectroscopy 207
   JOHN SUMMERSCALES and DAVID SHORT

Index 271

List of Contributors

R. D. ADAMS
Department of Mechanical Engineering, University of Bristol, Bristol BS8 1TR, UK

M. ARRINGTON
Speedtronics Ltd, 18 Pennway, Somersham, Huntingdon, Cambridgeshire PE17 3JJ, UK

A. F. BLOM
Aeronautical Research Institute of Sweden, PO Box 11021, S-161 11 Bromma, Sweden

P. CAWLEY
Department of Mechanical Engineering, Imperial College, London SW7 2BX, UK

P. A. GRADIN
AS Veritas Research, PO Box 300, N-1322 Høvik, Norway

JAMES MCKELVIE
Department of Mechanics of Materials, University of Strathclyde, James Weir Building, 75 Montrose Street, Glasgow G1 1XJ, UK

K. E. PUTTICK
Department of Physics, University of Surrey, Guildford, Surrey GU2 5XH, UK

DAVID SHORT
Department of Mechanical Engineering, Plymouth Polytechnic, Drake Circus, Plymouth PL4 8AA, UK

JOHN SUMMERSCALES
Royal Naval Engineering College, Manadon, Plymouth PL5 3AQ, UK

COLIN A. WALKER
Department of Mechanics of Materials, University of Strathclyde, James Weir Building, 75 Montrose Street, Glasgow G1 1XJ, UK

Introduction

In the time since the Second World War, fibre-reinforced plastics (FRP) have developed from a laboratory curiosity to become an industry which delivered 976 600 tonnes of composites in the USA alone during 1984, a record year (*Composites*, July 1985). New highly stressed components of FRP construction are regularly announced in most areas of structural design. The bulk of these materials use glass fibres as the reinforcement, but the use of carbon and aramid fibres is expanding rapidly. The technology of these materials has been well reviewed in a number of recent books [1–8] to which readers unfamiliar with composite materials are referred.

Over the corresponding period aspects of the physics and chemistry of materials have been allied with the rapid advances in electronics and computing to produce the new discipline of non-destructive testing (NDT). Abbreviations also used include NDE (examination/evaluation) and NDI (inspection). The subject of non-destructive testing has an extensive literature, of which the excellent comprehensive reviews [9–17], Royal Society discussion meeting [18] and the recent World Conference in Las Vegas [19] serve to outline the state-of-the-art.

This volume is the first part of a comprehensive in-depth review of the state-of-the-art of the use of non-destructive testing techniques as applied to the complete range of fibre-reinforced plastics. The literature for this field of interest is distributed across a diverse selection of journals ranging from the pure sciences to engineering. A number of conferences with a theme of the NDT of FRP have recently taken place [20–26]. Several short reviews of the subject also exist [27–52].

This volume considers most of the major techniques currently used for

NDT of composites: radiography, acoustic emission, thermal, optical and vibration methods. A technique which has fallen from favour, corona discharge, is then briefly reviewed. Finally, several chemical methods are considered because of their future potential as sensitive NDT techniques for polymer composites.

The next volume will primarily consider ultrasonic methods, which are the most common techniques currently practised for NDT of composites. Other secondary techniques will then be reviewed including eddy current, dielectric properties, microwave, electrical and magnetic means, fracto-emission and the Russell effect. Image and signal processing, including computer tomography, will also be covered. It is inevitable in works of this kind that some techniques of potential value will be omitted, either because they have not been adequately evaluated for the application (e.g. applied potential tomography [53]) or because they do not easily fall into the chapter structure of the text and are deemed to be of limited value (e.g. the vacuum displacement transducer [54], coefficient of restitution [54] or hardness testing [55–66], especially where they may inflict damage on the test-piece or component.

Composite materials must be regarded as very different media from metals, when considering which NDT methods are appropriate. Generally the reinforced plastics have poor electrical conductivity, low thermal conductivity, high acoustic attenuation and significant anisotropy of the mechanical and physical properties. The life of a metal component is determined by the nucleation and growth of cracks or damage in the material. The development of linear elastic fracture mechanics is often adequate as a basis for the definition of the size of subcritical flaws which must be identified.

However, a fibre-reinforced plastic is a heterogeneous medium which can contain multiple defect geometries. No single failure model can adequately describe the level of damage which is critical. A multiplicity of models have been developed to describe the various failure possibilities: interlaminar debonding, matrix degradation, fibre fracture and fibre–matrix interface separation. These in turn may be caused by improper cure, fibre misalignment, inclusions, poor reinforcement distribution, machining damage, fastener fretting and environmental degradation.

Non-destructive evaluation has three major functions for research, development and applications testing in composite materials. They are:

(a) initial inspection of test specimens and confirmation of the structural integrity of new components;

(b) monitoring sample tests in progress, or components subjected to service loads;

(c) analysing test results after failure, or proof loading of components during their service life.

In the case of industrial products, the design and production have always been considered the pre-eminent engineering challenges, with inspection, testing and defect diagnosis relegated to very subsidiary roles. However, there is an increasing awareness of the potential for in-service monitoring by NDT methods to improve the reliability of components in service and permit prolonged safe utilisation of the component.

This book, and its companion second volume, are therefore presented in an attempt to comprehensively review all aspects of the non-destructive testing of fibre-reinforced plastics, and hence to provide a convenient point of reference. Apart from a single volume by Potapov and Pekker [67], which is in Russian with no known translation, this is believed to be the only such text available to date.

JOHN SUMMERSCALES

REFERENCES

1. M. Grayson, *Encyclopaedia of Composite Materials and Components*, Wiley, New York & Chichester, 1983, ISBN 0-471-87357-8.
2. J. C. Halpin, *Primer on Composite Materials: Analysis*, Technomic, Lancaster, PA, 1984, ISBN 0-87762-349-X.
3. N. L. Hancox, *Fibre Composite Hybrid Materials*, Applied Science, Barking, UK, 1981, ISBN 0-85334-928-2.
4. D. Hull, *An Introduction to Composite Materials*, Cambridge University Press, Cambridge, UK, 1981, ISBN 0-521-23991-5.
5. G. Lubin, *Handbook of Composites*, SPE/van Nostrand Reinhold, New York, 1982, ISBN 0-442-24897-0.
6. M. R. Piggott, *Load Bearing Fibre Composites*, Pergamon, Oxford, 1980, ISBN 0-08-024230-8.
7. M. M. Schwartz, *Composite Materials Handbook*, McGraw-Hill, New York, 1984, ISBN 0-07-055743-8.
8. R. P. Sheldon, *Composite Polymeric Materials*, Applied Science, Barking, UK, 1982, ISBN 0-85334-129-X.
9. R. S. Sharpe (editor), *Research Techniques in Nondestructive Testing*, Vol. 1, Academic Press, 1970, ISBN 0-12-639050-9.
10. *Ibid.*, Vol. 2, Academic Press, 1973, ISBN 0-12-639052-5.
11. *Ibid.*, Vol. 3, Academic Press, 1977, ISBN 0-12-639053-3.
12. *Ibid.*, Vol. 4, Academic Press, 1980, ISBN 0-12-639054-1.

13. *Ibid.*, Vol. 5, Academic Press, 1982, ISBN 0-12-639055-X.
14. *Ibid.*, Vol. 6, Academic Press, 1982, ISBN 0-12-639056-8.
15. *Ibid.*, Vol. 7, Academic Press, 1984, ISBN 0-12-639057-6.
16. *Ibid.*, Vol. 8, Academic Press, 1985, ISBN 0-12-639058-4.
17. R. B. Thompson and D. O. Thompson, Ultrasonics in nondestructive evaluation, *Proceedings IEEE*, December 1985, **73**(12), 1716–1755.
18. E. A. Ash and C. B. Scruby (organisers), *Proceedings of the Royal Society of London discussion meeting: 'Novel techniques of non-destructive examination and validation'*, London, 9–10 July 1985, *Philosophical Transactions of the Royal Society of London*, 1986, **A320**(1554), 159–378.
19. *Proceedings of the 11th World Conference on Nondestructive Testing*, Las Vegas, 3–8 November 1985.
20. AGARD Structures and Materials Panel, *Nondestructive inspection relationships to aircraft design & materials, AGARD Conference Proceedings CP-234*, 1978, Proceedings of the 45th Panel Meeting, Voss, Norway, September 1977.
21. R. B. Pipes (editor), *Nondestructive evaluation and flaw criticality for composite materials*, Proceedings of a meeting, Philadelphia, October 1978, ASTM Special Technical Publication STP-696, December 1979.
22. K. L. Reifsnider (editor), *Damage in composite materials: basic mechanisms, accumulation, tolerance and characterisation*, Proceedings of a meeting, Bal Harbour, November 1980, ASTM Special Technical Publication STP-775, June 1982.
23. *Techniques for the characterisation of composite materials*, Proceedings of a Critical Review meeting, ONR, Massachusetts Inst. Tech., June 1981; AMMRC-MS-82-3, May 1982, AD A116 733.
24. C. E. Browning (editor), *Composite materials: quality assurance and processing*, Proceedings of a meeting, St Louis, October 1981, ASTM Special Technical Publication STP-797, February 1983.
25. T. Feest (editor), *Testing, evaluation and quality control of composites*, Proceedings of an International Conference, Guildford, September 1983, Butterworth Scientific, Sevenoaks, UK, 1983.
26. F. L. Matthews (guest editor), Defects in composites: detection and significance, Proceedings of a meeting, London, April 1985, *Composite Structures*, 1985, **3**(3–4), special issue.
27. M. A. Hamstad, A review: acoustic emission, a tool for composite-material studies, *Experimental Mechanics*, March 1986, **26**(1), 7–13.
28. W. N. Reynolds, Nondestructive examination of composite materials – a survey of European Literature; UKAEA Harwell Report AERE G-1757, May 1980, TRC T81-4170, AD A086 165; UKAEA Harwell Report AD A100 454, 1981, AMMRC-TR-81-24, USAAVRADCOM-TR-81-F-6; *Materials and Design*, December/January 1985, **5**(6), 256–270.
29. W. N. Reynolds, Nondestructive testing of fibre-reinforced composite materials, *SAMPE Quarterly*, July 1985, **16**(4), 1–16.
30. I. G. Scott and C. M. Scala, *NDI of composite materials*, Australian Aeronautical Research Laboratory, ARL-MAT-TM-379, July 1981, AD A106 278.

31. I. G. Scott and C. M. Scala, A review of nondestructive testing of composite materials, *NDT International*, April 1982, **15**(2), 75–86.
32. D. H. Kaelble, *Quality control and nondestructive evaluation techniques for composites. Part 1: Overview of characterization techniques for composite reliability*; USAAVRADCOM-TR-82-F-3, May 1982, AMMRC-TR-82-36, AD A118 410, N83-11508.
33. J. L. Koenig, *QC & NDE techniques for composites. Part 2: Physicochemical characterisation techniques — calorimetric analysis, dynamic mechanical analysis, infrared and Raman spectroscopy and dielectric analysis*, USAAVRADCOM-TR-83-F-6, May 1983, AMMRC-TR-83-24, AD A131 038, N84-10200.
34. *QC & NDE techniques for composites. Part 3: Liquid chromatography* (not published to date).
35. F. P. Alberti, *QC & NDE techniques for composites. Part 4: Radiography*, USAAVRADCOM-TR-82-F-4, June 1982, AMMRC-TR-82-40, AD A137 156.
36. *QC & NDE techniques for composites. Part 5: Ultrasonics* (not published to date).
37. M. A. Hamstad, *QC & NDE techniques for composites. Part 6: Acoustic emission*, USAAVRADCOM-TR-83-F-7, May 1983, AMMRC-TR-83-25, AD A132 621, N84-14245.
38. E. G. Henneke and K. L. Reifsnider, *QC & NDE techniques for composites. Part 7: Thermography*, USAAVRADCOM-TR-82-F-5, March 1982, AMMRC-TR-82-18, AD A114 392.
39. *QC & NDE techniques for composites. Part 8: Applications to the manufacture of composite main rotorblade* (not published to date).
40. D. H. Kaelble and R. J. Shuford, Composite characterisation techniques: overview, *Mantech Journal*, 1985, **10**(2), 17–26.
41. J. L. Koenig, Composite characterisation techniques: physicochemical, *Mantech Journal*, 1985, **10**(2), 27–36.
42. F. P. Alberti, Composite characterisation techniques: radiography, *Mantech Journal*, 1985, **10**(2), 37–45.
43. S. Serabian, Composite characterisation techniques: ultrasonics, *Mantech Journal*, 1985, **10**(3), 11–23.
44. M. A. Hamstad, Composite characterisation techniques: acoustic emission, *Mantech Journal*, 1985, **10**(3), 24–32.
45. K. L. Reifsnider and E. G. Henneke, Composite characterisation techniques: thermography, *Mantech Journal*, 1985, **10**(3), 3–10.
46. AGARD/W. L. Shelton, NDI of composite materials, AGARD-AG-201, *Nondestructive Inspection Practices*, Vol. 2, October 1975, pp. 579–592.
47. C. M. Teller, G. L. Burkhardt and G. A. Matzkanin, *Nondestructive evaluation of carbon–carbon composites: a state-of-the-art survey*, Southwest Research Institute, AD B032 321L, November 1978.
48. M. L. Phelps, *Assessment of the state-of-the-art of in-service inspection methods for graphite–epoxy composite structures on commercial transport aircraft*, NASA-CR-158 969, January 1979, N79-17252, also N82-12142, November 1981.

49. G. A. Matzkanin, G. L. Burkhardt and C. M. Teller, *Nondestructive evaluation of fibre reinforced epoxy composites: a state-of-the-art survey*, Southwest Research Institute, April 1979, AD A071 973.
50. E. G. Henneke and J. C. Duke, *A review of the state-of-the-art of nondestructive evaluation of advanced composites materials*, Virginia Polytechnic Institute, September 1979, ORNL/sub-79-13673/1 [DE83 000983].
51. A. Vary, A review of issues and strategies in nondestructive evaluation of fibre reinforced structural composites, *11th National SAMPE Technical Conference*, Boston, November 1979, pp. 166–177.
52. G. A. Matzkanin, *Nondestructive evaluation of fibre reinforced composites: a state-of-the-art survey*. Volume 1: NDE of graphite fibre reinforced plastic composites. Part 1: Radiography and ultrasonics, Southwest Research Institute, March 1982, NTIAC-82-1.
53. B. H. Brown, Applied potential tomography – looking inside the human body and other conducting objects, *Physics Bulletin*, March 1986, **37**(3), 109–112.
54. J. D. Brannon and R. J. McNicholls, *Nondestructive testing methods for shelter panels*, Army Material Command report USAMC-ITC-02-08-75-002, April 1975, AD A009 286.
55. F. N. Cogswell, Microstructure and properties of thermoplastic aromatic polymer composites, *SAMPE Quarterly*, 1983, **14**(4), 33–37.
56. D. Short, A. W. Stankus and J. Summerscales, Woven glass-fibre reinforced polyester-resin composites exposed to the marine environment, *Proc. 1st International Conference on Testing, Evaluation and Quality Control of Composites*, Guildford, September 1983, pp. 212–220, Paper 21.
57. Shaw Ming Lee, Double torsion fracture toughness test for evaluating transverse cracking in composites, *Journal of Materials Science Letters*, December 1982, **1**(12), 511–515.
58. M. A. Rizzi and M. R. Kearney, Large injection moulded polyester premix parts with increased toughness, *Proc. 37th Annual Reinforced Plastics/Composites Institute Conference*, SPI, Washington, DC, January 1982, Paper 13-D.
59. S. S. Parikh and P. G. Daugherty, Effect of pigment types and percentages on mechanical properties of polystyrene fibreglass reinforced thermoplastics, *Proc. 40th Annual Technical Conf*, SPE, San Francisco, May 1982, pp. 405–408.
60. A. Sternfield, New block copolymers: impact-modified nylons offer more options, *Modern Plastics International*, 1982, **12**(2), 22–25.
61. P. Michel, Nouveau procédé de fabrication de polyesters insaturés expansés et applications au renforcement par fibres de verre, *Plast. Renf. Fibres Verre Tex.*, 1981, **21**(8), 21–25.
62. E. S. W. Kong, Tg annealing studies of advanced epoxy-matrix graphite-fibre reinforced composites, *Journal of Applied Physics* (USA), 1981, **52**(10), 5921–5925.
63. G. Ehnert, Further prospect of market requirements and SMC developments for automotive applications in Europe, *Proc. 35th Annual Reinforced Plastics/Composites Institute Conference*, SPI, New Orleans, February 1980, Paper 8-D.

64. T. O. Bautista, The role of synthetic veil in the wear factor of corrosion resistant laminates, *Proc. 35th Annual Reinforced Plastics/Composites Institute Conference*, SPI, New Orleans, February 1980, Paper 5–B.
65. P. Salwiczek, Les polyurethannes thermoplastiques caoutchoutiques, situation actuelle et perspectives, *Les Caoutchoucs Thermoplastiques dans les Années 80*, Journée AFICEP/SPE-France, Paris, 1980, Paper D1.
66. *Glass reinforced plastics: measurement of hardness by means of a Barcol impressor*, British Standard BS 2782 Part 10, method 1001, 31 August 1977; Europäische Norm EN59 Edition 1, March 1977.
67. A. I. Potapov and F. P. Pekker, *Nondestructive testing of composite materials*, Mashinostroenie, Leningrad, 1977, in Russian.

Chapter 1

Radiography

A. F. BLOM
Aeronautical Research Institute of Sweden, Bromma, Sweden
and
P. A. GRADIN
AS Veritas Research, Høvik, Norway

1.1. INTRODUCTION

It was in 1895 that Wilhelm Conrad Röntgen discovered that a certain type of radiation, caused by electric discharges in rarefied gases, was able to penetrate, for example, paper, wood, the human body (in fact the first human radiograph was made of Mrs Röntgen's hand) and light metals. The following year Becquerel discovered that certain uranium bearing ores emitted a similar but more penetrating type of radiation, the so-called gamma rays.

During the First World War, the radiation discovered by Röntgen, or the X-rays as he called it, was used to examine, for example, wooden aeroplane propellers for cracks. Since that time X- and gamma rays have been used to a growing extent in the most diverse applications.

This chapter will present, first, a rather brief introduction to the basic physical principles upon which the use of radiography is based and, second, a summary of the different applications of non-destructive testing and evaluation of fibrous composites.

1.2. PRINCIPLES

For a very good and thorough presentation of the basic principles in radiography, the study of references [1]–[4] is strongly recommended. In fact, the following sections are very much a summary of those references.

To produce an X- or gamma ray radiograph, a beam of the radiation is directed towards the object that is to be investigated, and the differences in radiation after passage of the object are recorded on a special type of film. The differences in intensity will be caused by differences in thickness and density of the object. In Fig. 1 the principle of the radiographic process is shown.

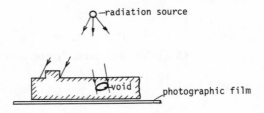

FIG. 1. Principle of radiographic process.

When using a neutron source instead of an X-ray or gamma ray source, a photographic film sensitive to X- or gamma rays can not be used directly. In such a case a screen is used, which when excited by neutrons emits radiation that affects the X-ray film. When the film is processed one will obtain a 'picture' of the internal structure of the object, the interpretation of which will in general demand both experience and skill.

1.2.1. X-rays

X-rays are produced when electrons in vacuum are accelerated to a high velocity in order to hit a target (usually made of tungsten); see Fig. 2.

The electrons are produced by heating of the filament. A high voltage (tube voltage) between the cathode and anode causes the electrons to accelerate and strike the target, thereby causing the emission of X-rays. The flow of electrons is, through the action of the focusing cup, concentrated in a small region of the target called the focal spot, the size of which is an important parameter in X-ray radiography. Different procedures for determining the focal spot size exist (cf. [5]). In Fig. 2, the X-ray

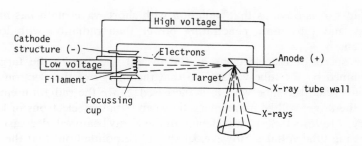

FIG. 2. Schematic description of an X-ray tube (from Ref. [1], reproduced from *Materials Evaluation* with the permission of the American Society for Nondestructive Testing and the American Society for Metals, 1973).

tube produces a cone of radiation, but there also exist X-ray tubes that produce a radiation field extending 360° around the tube.

For tube voltages up to about 400 kV, the tubes are constructed so that the electrons generated at the filament follow a linear path toward the target. In the voltage range 1–30 MV some X-ray tubes are constructed in such a way that the electrons are accelerated in a spiral path. This is achieved by applying a strong magnetic field normal to the electron path.

X-ray tubes used for industrial applications are classified by the tube voltage they produce, the classification being done according to:

   5 kV–50 kV   low voltage
   50 kV–300 kV medium voltage
   300 kV–1 MV   high voltage
   1 MV–30 MV   super voltage

When testing fibre composites, the low voltage and the lower part of the medium voltage region are normally used.

Before discussing the influence of, for example, the tube voltage on the X-rays, a formula for determining the energy of electromagnetic radiation will be given:

$$E = hf = hc/\lambda \qquad (1)$$

where E is the energy, f the frequency, h Planck's constant, c the speed of light and λ the wavelength of the radiation. Since in eqn (1) both h and c are constants, it is obvious that radiation with short wavelength has higher energy and thus more penetrating power than radiation with longer wavelength.

The speed of the electrons at the instant they strike the target is determined by the tube voltage, so that the energy of the electrons will increase when the tube voltage is increased. Since the radiation emitted from the target will be more energetic when struck by electrons of larger energy, it follows that the wavelength of the X-ray beam will decrease with increasing tube voltage. However, it should be pointed out that the radiation emitted from the target does not consist of a single wavelength, but consists of such a large number of wavelengths that it can be considered as a continuous spectrum. The longest wavelengths (with little penetrating power) in this spectrum are absorbed by the tube walls and the tube port. This is referred to as the inherent filtration of the tube head.

To increase the intensity of the radiation, one could increase the energy of the electrons reaching the target (i.e. increase the tube voltage) or one could increase the number of electrons that reach the target per time unit. However, the major purpose of increasing the tube voltage is to increase the penetration power, and it is the flow of electrons (the tube current) that is varied when the intensity is to be changed.

1.2.2. Gamma rays

As was mentioned before, X-rays and gamma rays are similar in nature although they have completely different sources. The sources of gamma rays are certain radioactive isotopes, and one advantage of using this type of radiation is obvious, namely independence from electrical power. Other advantages are simplicity of apparatus, convenient portability and compactness of the radiation source. Gamma ray sources do not, as X-ray tubes, produce a continuous spectrum, but emit only a few discrete wavelengths. These wavelengths (which determine the energy and hence the penetrating power as in eqn (1)) cannot be adjusted, but are determined by the radioactive source.

A unit for specifying the energy of gamma radiation (and also high and supervoltage X-rays) is eV (electron volts). 1 eV is the kinetic energy of an electron moving through a potential difference of one volt. Multiples of eV often used are keV and MeV (thousand and million electron volts, respectively). It is not possible to say that the penetrating power of a 500 keV gamma ray source is equal to that of a 500 kV X-ray tube. The shortest wavelength in the X-ray spectrum is about equal in penetrating power to the gamma ray, but the rest of the wavelengths will be far less penetrating. As an example, gamma radiation from cobalt 60, with two wavelengths of energy 1.17 and 1.33 MeV, is equivalent to the radiation from a 2 MV X-ray tube. For a radioactive source, the intensity is directly related to the number of disintegrations during a time unit. The activity equalling 3.7×10^{10} disintegrations per second is called 1 curie.

If the energy from all disintegrations was radiated out from the source, then the intensity of the radiation would be directly proportional to the activity (number of disintegrations). However, and this is valid especially for large radiation sources, some of the energy will be absorbed by the source itself; this is referred to as self-absorption.

By multiplying the source activity by the so-called gamma ray dosage factor (a factor specific for the type of source used) one will obtain the intensity in Röntgen (neglecting the self-absorption). One disadvantage with the use of gamma ray sources is that the activity or intensity cannot be adjusted but is given by the source used and the time elapsed. The activity is halved after a time called the half-life, which for iridium 192 is 70 days and for cobalt 60 is 5.3 years.

1.2.3. Neutron radiography

As the name implies, neutron radiography uses a flux of neutrons (subatomic particles from the nucleus of an atom, carrying no electric charge). The absorption of X- and gamma rays is in general increased with increasing atomic number of the material under inspection. This is not true in the case of neutrons, which are most readily absorbed by materials containing hydrogen (such as polymeric matrices) and least absorbed by materials with high atomic numbers (exceptions to this rule do exist).

As was mentioned before, neutrons do not affect X-ray film, so when such film is to be used a screen is placed in direct contact with the film. This screen, when excited by neutrons, emits radiation to which the film is sensitive. In a so-called autoradiographic technique the screen only is exposed to the neutrons and then immediately placed upon the film in order to produce a radiograph.

Different sources for neutrons exist. One (perhaps obvious) source is nuclear reactors, where the neutrons are supplied through a port extending into the reactor. There also exist electronic sources for producing fast neutrons, which have to be slowed down (moderated) before use. Finally, there exist isotope sources such as californium 252, which produce fast neutrons that also have to be moderated before use in radiography.

1.2.4. Use of screens
Screens are used in order to intensify the radiation transmitted through the detail that is to be radiographed. The screens are of two types, fluorescent intensifying screens and lead screens.

1.2.4.1. *Fluorescent screens*
Some chemicals such as barium lead sulphate and calcium tungstate have the property that they emit visible light when exposed to X- or gamma rays. This is used in radiographic applications in the following way: the film is placed between and in contact with two screens so that the film will be affected both by visible light and by X-rays, so that the total effect will be the sum of these two effects. It is possible to obtain an intensifying factor typically in the order of 10 to 60 times.

There are, however, some drawbacks in using fluorescent screens. One is that the screen spreads the light and produces a radiograph with fuzzy contours. This is because, even if the X-rays are very well-defined when exciting the screen, the light from the screen will be emitted in all directions. Other drawbacks are that not only primary but also scattered radiation is intensified; further, fluorescent screens absorb scattered radiation as do lead screens.

1.2.4.2. *Lead screens*
These are in fact thin lead foils placed on both sides and in close contact with the X-ray film. The foil nearer to the radiation source typically has a thickness of 0.02–0.15 mm while the foil on the back of the screen is usually about twice as thick. The function of the back screen is to absorb scattered radiation, while that of the front screen is to generate electrons and secondary radiation in order to increase the intensity transmitted through the radiographed object, thereby reducing the exposure time.

1.2.5. Quality of radiographs
In this section will be presented some of the factors that affect the quality of a radiograph. Some of the geometric factors affecting the radiograph will be explained in terms of visible light. Consider a light L shining on an opaque object O, forming a shadow on a white card C (Fig. 3). If the object is not placed directly on the card but some distance away, the shadow will be an enlarged picture of the object, the degree of enlargement being given by:

$$\frac{S_i}{S_o} = \frac{D_i}{D_o} \tag{2}$$

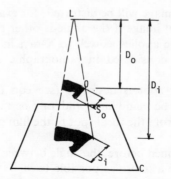

FIG. 3. Degree of enlargement.

In Fig. 3, L is assumed to be a point source. This is not true for an X-ray source, since the focal spot has finite dimensions which will give rise to so-called geometric fuzziness. This is illustrated in Fig. 4.

The geometric fuzziness, U_g, is easily obtained from Fig. 4 as:

$$U_g = F \cdot \frac{t}{D_o} \tag{3}$$

where F, t and D_o are defined in Fig. 4. Another factor which might affect the image is the angle between the central ray and the film plane. If this

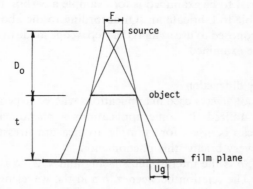

FIG. 4. Geometric fuzziness (from Ref. [2], reproduced from *Materials Evaluation* with the permission of the American Society for Nondestructive Testing and the American Society for Metals, 1973).

angle is not 90° the image will be distorted; for example, a circular object will have an elliptical image if the angle is other than 90°.

What is said above applies as well to X-rays as to gamma rays. Some general rules to be considered in radiographic applications are given below.

(1) A small focal spot is desirable (see eqn (3)). If this cannot be achieved, D_o should be made large and/or t be made small.

(2) The distance from the focal spot to the film plane should be as large as possible.

(3) As was mentioned before, the angle between the central ray and the film plane should be as close as possible to 90°.

An important parameter in radiography is the exposure. In the case of an X-ray tube, with tube current m acting during a time t, the exposure D is defined by:

$$D = mt \qquad (4)$$

the unit for D being, for example, mA s (milliampere-seconds). The exposure for a gamma ray source will also be defined according to eqn (4), but m will be the source activity in curies (Ci), a suitable unit for D in this case being curie-minutes. To assess the quality of a radiograph a so-called image quality indicator (IQI) is often used; IQIs are prescribed in various standards. Most commonly, an IQI consists of a set of wires with different diameters and made of the same material as the object to be radiographed. A measure of the radiographic quality is the ratio of the diameter of the smallest wire that can be seen on the radiograph to the object thickness. If the material to be examined is for example a carbon fibre composite, it is not possible to fabricate an IQI according to the above. In ref. [6] it is therefore proposed to use as an IQI a step wedge made of the same material that is to be examined.

1.2.6. X-ray diffraction

X-rays are not always used in applications where the penetrating ability of the rays is utilized. In some applications a phenomenon called X-ray diffraction can be used, for example, to measure stresses. Figure 5 helps to explain, very briefly, this phenomenon.

In Fig. 5, b is the interatomic spacing and $\pi/2 - \alpha$ is Bragg's angle of diffraction. The relation between α, b and the wavelength λ is

$$\lambda = 2b \cos \alpha \qquad (5)$$

It is possible to measure α quite accurately (Fig. 6). When θ is varied, the

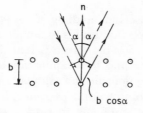

FIG. 5. Two parallel X-ray beams diffracted from two atomic planes.

intensity of the diffracted X-rays will show a maximum for a certain value θ_0 of θ. The relation between θ_0 and α is given by

$$\theta_0 = \frac{\pi}{2} - \alpha \tag{6}$$

Equation (5) can be used for strain measurements in the following way. Assume that the interatomic spacing is b when the structure is unloaded. From (5) is obtained:

$$b = \lambda/(2\cos\alpha) \tag{7}$$

Now, due to loading, b will be changed by an amount Δb (assumed small). From eqn (7) follows:

$$\Delta b = \lambda \sin\alpha \Delta\alpha/(2\cos^2\alpha) \tag{8}$$

The strain ε_n in the n-direction (see Fig. 5) is now obtained as:

$$\varepsilon_n = \Delta b/b = \tan\alpha\, \Delta\alpha \tag{9}$$

where $\Delta\alpha$ is the change in α due to loading.

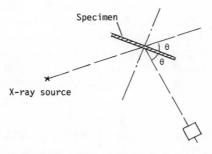

FIG. 6. Schematic representation of an X-ray diffractometer.

1.2.7. Filmless techniques

One way of avoiding the use and handling of X-ray film when utilizing X-ray radiography as an inspection tool is to use a fluoroscopic screen. One advantage of using a fluoroscopic screen is that the image obtained appears in real time, which permits the examination of moving objects. Disadvantages are that the equipment and especially the viewing glass must be well shielded in order to protect the personnel from receiving excessive radiation. Furthermore, it is difficult to work at the viewing screen for more than 30 minutes due to eye fatigue, because of the cyclical nature of the exposing radiation.

Another filmless technique is called Xeroradiography. In this technique a plate is used, the surface of which becomes radiation sensitive when charged by static electricity. At this instant the plate must be protected from light. Due to the ionizing nature of X-ray radiation, the surface of the plate will be discharged according to the amount of radiation that is incident on the plate. To develop the image, a powder consisting of small plastic particles, charged in an opposite sense to the plate, is used. The density of particles on the plate will depend on the electric charge and also on the amount of radiation received, and in this way an image is formed. To make the image permanent, the plastic particles are transferred over to a special type of paper (by pressing the paper onto the plate) and are then fixed by heating the paper. To obtain a positive image, a light coloured paper is used together with dark coloured particles, and vice versa if a negative image is wanted.

1.2.8. Computer-assisted radiography

By using a computer in connection with radiography, it is possible to process the image in various ways. A system for computer based radiography usually consists of three units or subsystems: an image acquisition system, an image conditioning system and finally an image processing system. The signal from the test object is sensed in some way and the resulting image is digitized and stored in a memory, the data of which then can be processed by use of a computer. One operation could for example be to increase the contrast over a region of the object. Reference [7] describes a system developed for computer assisted radiography of composite materials.

1.3. APPLICATION TO FIBRE-REINFORCED COMPOSITES

As already discussed in the previous section, there exists a large number of different radiographic techniques for non-destructive testing and evaluation of fibre-reinforced plastics. In this section we will discuss applications of some of these techniques to the non-destructive study of different property variations, defects and other damage in fibre-reinforced composites. General reviews of non-destructive testing of composites, including discussions of radiographic techniques, are presented in, for example, Refs [6, 8–11].

1.3.1. Volume fractions of fibre and resin

The relative volume fractions of fibre and resin may be established by measuring either the fibre or the resin content. However, consideration must be given to the relative mass absorption coefficients of the fibre and matrix in the studied composite. Martin [12] used both low energy X-rays and thermal neutrons to measure the resin content in carbon/epoxy composites. He found that a 1% change in resin content corresponds to a 2.6% change in the thermal neutron mass absorption coefficient, whereas the equivalent change for X-rays is only 0.7%. On the basis of these results, Martin concluded that the use of X-ray absorption techniques for measuring composite resin content is not feasible. He also concluded, on the basis of the experimental measurements, that the neutron film technique used was not sensitive enough for practical measurements of resin content. However, it was suggested that neutron gauging techniques, which are more accurate than film techniques for hydrocarbons such as epoxy, should be able to detect changes in the resin content of $\pm 1\%$. Further experimental work is required to substantiate this suggestion.

For a glassfibre composite with a mass absorption coefficient ratio of 20 between glass and the unpigmented polyester, Förli and Torp [13] were able to determine the total glass content from film density measurements.

1.3.2. Fibre alignment and fibre flaws

It has been shown by Crane *et al.* [14] that the addition of boron fibres to the edges of carbon/epoxy prepreg tapes permits radiographic inspection of both the distribution and integrity of fibres on a tape-by-tape level within a component. The boron fibre is coated with a dilute epoxy and then dusted with a fluorescent dye powder which makes the fibre appear as a bright yellow line against the black background of the carbon tape. It was shown [14] that by using boron fibres with a failure strain approximately equal to that of the composite reinforcing fibre, it was possible to estimate the location and extent of foreign damage by noting the number of broken boron fibres at the actual area. In this way a good estimate of the depth of internal fibre breakage was obtained for specimens with a radius of curvature greater than about 6.5 mm, which is a limitation in the use of boron fibres due to their strength and stiffness properties.

1.3.3. Use of opaque penetrants

In fatigue loading of composites a large number of failure mechanisms, such as matrix cracking, fibre breakage, debonding and delamination, may occur simultaneously depending on the material system, the local geometry and the mode of loading [15, 16]. The one failure mechanism currently of most concern is delamination, which may propagate between individual plies until fracture occurs. Conventional X-ray techniques cannot easily be applied to delamination monitoring in carbon/epoxy laminates due to poor contrast. The most effective, and now very common, way to detect internal damage such as delaminations involves the use of tetrabromoethane (TBE) or diiodobutane (DIB) as contrasting fluids. These fluids are commonly used by reason of their penetrating and image properties [17, 18]. Both these penetrants are halogenated hydrocarbons and therefore require special precautions in handling, use and storage [17, 18]. In fact, although S-tetrabromoethane is more opaque to X-rays than 1, 4 diiodobutane it is not recommended for industrial applications as it is listed as a severe poison and potent mutagen, whereas DIB is classified only as an irritant [8]. Although both TBE and DIB are commonly used in many research projects, there has recently been much interest in finding another less harmful but still useful opaque additive. Rummel et al. [19] studied a number of candidate X-ray opaque materials and found that a zinc iodide based solution (for which no special handling precautions have been identified for general use) was capable of equalling the sensitivity of the halogenated organic compounds. The developed solution consisted of 60 g zinc iodide (ZnI_2) in 10 ml of water (H_2O), 10 ml of isopropyl alcohol ($CH_3CHOHCH_3$) and 1 ml of Kodak 'Photo Flo 600' which is a linear alcohol alkoxylate used as a wetting agent to reduce the surface tension of the solution. The isopropyl alcohol was used to improve the penetrant properties by giving the solution both polar and non-polar solvent properties. The zinc iodide solution used was not saturated and thus it was concluded in ref. [19] that the sensitivity may be further improved by varying the concentration. The exposure of carbon/epoxy materials to TBE, DBI or zinc iodide solution was shown in [19] to have no significant influence on carbon/epoxy material properties, either at room temperature or at elevated temperatures.

Using TBE-enhanced X-ray radiography, Mohlin et al. [20, 21] studied delamination growth in notched carbon/epoxy laminates subjected to compressive fatigue loading. Specimens of Fiberite T300/1034E with a stacking sequence $[\pm45/0_2/\pm45/90/0_3/\pm45/0_2]_s$ were fatigue loaded in pure compression ($R = \sigma_{min}/\sigma_{max} = -\infty$) at a cyclic frequency of 6–10 Hz. At certain intervals of fatigue cycles, the specimens were removed from the testing machine and X-rayed using a TBE-enhanced technique to provide the necessary contrast. The TBE was applied to the surfaces of the specimens around the hole and the edges for about 30 min to give the fluid sufficient time to penetrate into the damaged regions. After it had fully saturated the damaged regions, the excess TBE was removed from the surfaces of the specimens with absorbent towels. The specimens were then laid directly on top of a Curix RP1 X-ray film and X-rayed. The X-ray head was located at a distance of 75 cm from the X-ray table. The energy used was 60 kV and 200 mA and the exposure time was 0.45 s.

At all studied load amplitudes (48–88% of static compressive fracture load), matrix cracks in the 0° plies became visible at the tangent axial to the hole boundary after less than 10^3 cycles. After a large number of cycles, delamination initiated at the hole boundary perpendicular to the loading direction and grew in a stable manner.

Figure 7 shows X-ray pictures taken for an applied load level of 74% of the static compressive fracture load, $P_{sc}(= 42$ kN) for the laminate. Initially, delamination is observed on one side of the hole. The delamination grows in both transverse and longitudinal directions. After a larger number of cycles, delamination initiates at the reverse side of the hole. When the damage has reached a larger size, rapid growth occurs until the delamination reaches the edges of the sample. Further fatigue cycling results in multiple delamination as found by visual observation of the edge of the sample, until eventually complete failure occurs.

In order to quantify the delamination growth the area of the delaminated regions was estimated. Figure 8 shows the delaminated area as a function of the number of fatigue cycles for different load levels.

It was shown in Ref. [20] that the delaminated area of the specimen was approximately linearly related to an overall loss of longitudinal stiffness regardless of the applied load level. The experimental results obtained were used together with a three-dimensional finite element model to calculate the strain energy release rate for a delamination crack located between the surface angle plies ($\pm45°$) and the adjacent 0°−ply [21]. Realizing that multiple delamination occurs, Blom [22] performed an analytical fracture mechanics analysis based on global energy considerations and managed to correlate, for various load levels, the growth rate

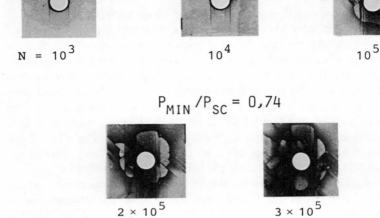

FIG. 7. X-ray pictures of delamination growth in a compressively loaded carbon/epoxy laminate (from Ref. [21]).

of the delaminations found with X-rays with the cyclic range in strain energy release rate.

Using opaque additives, most frequently TBE, several workers have performed similar studies to those described above on delamination growth due to fatigue loading, e.g. refs [23–30], but also on damage resulting from impact and creep, e.g. [31–33]. Most authors agree that penetrant-enhanced X-ray radiography gives detailed information on the nature and planar distribution of damage in fibre-reinforced composites. The technique is capable of detecting fibre fractures, matrix cracks and delaminations.

Conventional penetrant-enhanced radiographic techniques suffer mainly from two aspects.

Firstly, it is recognized that the only defects to be detected will be where capillary forces can bring the fluid into the damaged region. Thus, defects originating from free edges or notches will be readily observed, whereas buried delaminations will not. A means to overcome this problem was suggested by Ramkumar [27]. Small holes (about 0.10 mm in diameter) were drilled from the nearer surface to the delaminated interface using a 6 W laser beam. A hypodermic needle was then used to inject the opaque

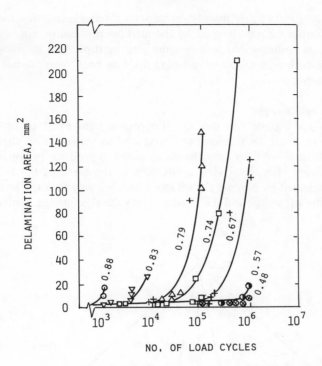

FIG. 8. Delamination growth during compressive fatigue at different load amplitudes. Parameter is P_{min}/P_{SC} (from Ref. [21]).

fluid (DIB) into the delaminated region. The small laser-drilled holes were assumed to have negligible detrimental effects on delamination growth.

Secondly, no information is obtained of the distribution of the damage through the thickness. To overcome this problem, several alternative techniques have been used. The simplest and most obvious way to study the progressive development of damage through the thickness of the studied specimen is to make additional radiographs with the edge of the specimens oriented perpendicular to the X-ray beam. Such a technique was successfully used by Black and Stinchcomb [30] to monitor delamination in thick, notched graphite/epoxy laminates. The extent of the observed damage in the X-ray radiographs was verified by sectioning the damaged laminates.

Other techniques to study through-thickness distribution of damage in fibre-reinforced plastics rely mainly on various three-dimensional radiographic techniques frequently used by the medical profession, e.g. stereo radiography, computer-aided tomography and multiple film laminography. Of these techniques, stereo radiography has been of most use for studies of composites.

1.3.4. Stereo radiography

Stereo X-ray radiography is the X-radiographic equivalent of optical stereography in which two images are produced as viewed from slightly different angles and are then optically recombined to produce an apparent three-dimensional view. This view exists only in the mind of the viewer and is accomplished by placing the images in such a way that the left eye can see only the left image and the right eye only the right image. Different

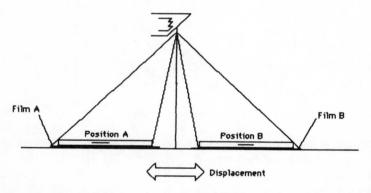

FIG. 9. Stereo radiographic image produced by object displacement method (from Ref. [19]).

techniques exist both to produce and to study stereo images. Rummel et al. [19] used an object displacement method, shown in Fig. 9, to produce stereo images.

Alternative methods would be (a) holding the object immobile while shifting the position of the X-ray tube, or (b) holding the X-ray source immobile as in Fig. 9 but, rather than displacing the object, tilting it with respect to the source axis. The latter method was used, for example, by Sendeckyj et al. [25]. For viewing the stereo images in the three-dimensional mode it is common to use commercially available stereoscopes which consist of a pair of convex lenses mounted in a hand-held frame which allows the images to be moved to compensate for the variation in the viewer's eyesight. A more convenient device was developed and used by Sendeckyj *et al.* [25] who built a unit, as shown in Fig. 10, which allowed for routine examination of specimens by several viewers, one at a time, without intervening adjustments.

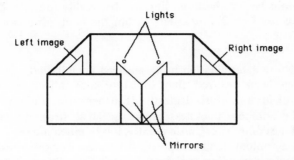

FIG. 10. Device for viewing X-ray stereo pair radiographs (from Ref. [25]).

It is reported [19, 25] that the use of stereo X-ray radiography provides detailed information of the through-thickness damage in composites. The technique is capable of locating a single ply within a laminate.

As already mentioned, there exist other techniques, commonly used for medical applications, that could be useful for through-thickness studies of composites. In computer-aided tomography the attenuation of an X-ray beam is recorded while the beam rotates, moving along different diameters through the object. Then the computer calculates images of cross-sections, which are related to density maps. The density resolution is of the order of 0.5 %. How this technique might be used in non-destructive analysis of composite materials is briefly discussed by More *et al.* [34]. We believe, however, that the use of stereo radiographic techniques, as discussed above, is fully adequate for damage studies of composite materials. The cost of computer-aided tomography probably prohibits its use in the field of composite materials.

1.3.5. Neutron radiographic inspection of bonded composite structures

In the inspection of bonded composite to metal joints or of adhesively bonded honeycomb structures, the X-ray attenuation coefficients of the different materials vary in such a way as to provide very little information regarding the quality of the bond. For such structures a neutron radiographic technique may be the only one providing a clear image of the bond or of the composite itself in the presence of metals [35–37]. Using mobile neutron radiographic systems, with the isotope californium-252, it has been shown that bond-line voids as small as 0.127 mm diameter in laminated carbon/epoxy specimens can be detected [36]. It is also reported that inclusions within the bond line will be readily found and that areas of excessive resin or fibre content within the composite can be located [36].

1.3.6. Moisture determination using positron annihilation

It has long been realized that moisture may affect the mechanical properties of fibre-reinforced plastics. Furthermore, the moisture profile through the laminate thickness can cause large stress gradients due to swelling of the composite. Consequently, it is of much interest to determine the moisture content in all composite structures. For small test specimens it is simple to determine moisture content and diffusion constants by drying the specimen, weighing it and then monitoring weight as a function of moisture absorption. The moisture profile may then be calculated for a larger panel using Fick's diffusion laws. For a real structure, however, there is a need to be able to measure, non-destructively, the moisture content and its depth distribution.

A technique which seems suitable for such measurements is based on the dependence of positron lifetime on the moisture content of the composite specimen [38–40]. The positron is the antiparticle of an electron. When a free positron, usually obtained from one of the artificially radioactive isotopes ^{22}Na, ^{58}Co, ^{64}Cu or ^{68}Ge, encounters an electron, the two annihilate each other with a lifetime of 125 picoseconds, emitting a pair of gamma rays. When a positron enters a molecular medium such as a polymer, there are at least two annihilation routes possible [38, 39]. There is one short-lifetime component related to direct annihilations with molecular electrons and to the self-annihilation of singlet positronium (parapositronium). The second longer lifetime, of the order of a few nanoseconds, is believed to relate to the annihilation of triplet positronium (orthopositronium) by the so-called pick-off process (see ref. [39]). This latter delayed positron annihilation includes the effects of interaction of the positronium with the surrounding media and responds sensitively to the changes in the properties of the medium produced by diffusing water.

Singh *et al.* [38, 39] used a conventional fast-slow coincidence system to measure positron lifetimes for different moisture contents in the studied materials. Positron lifetime is measured with respect to a reference time determined by the detection of a nuclear gamma ray emitted simultaneously with the positrons which were emitted from a 10 μCi ^{22}Na source.

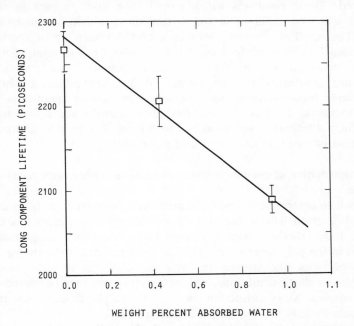

FIG. 11. Effect of moisture on positron lifetime for Narmco 5208/T300 carbon/epoxy (from Ref. [39]).

An example of the experimentally observed dependence of the long component lifetime on the degree of moisture content is shown in Fig. 11 for a Narmco 5208/T300 carbon/epoxy specimen. It is seen that the positron lifetime decreases linearly with the moisture content of the specimen. For industrial applications, the experimentally determined curve might be used as a calibration curve for determining the moisture content of objects fabricated from the same material.

It appears from Fig. 11 that the positron technique used by Singh *et al.* [38, 39] has sufficient sensitivity to monitor moisture content in conventional carbon/epoxy systems, even though the saturation weight percent of water is of the order of only 1%. Thus, for uniformly distributed moisture in these materials, continuous-energy positron-emitter radioactive sources, as discussed above, are quite adequate. However, for specimens having nonuniform moisture distribution, the positron annihilation characteristics will vary depending on where the positrons are thermalized. Those positrons which stop in the drier regions of the composite will have different lifetimes from those which stop in moist regions. The resulting spectrum will be a complex mixture of spectra characterized by different lifetimes. Thus, for investigating nonuniform moisture distributions in fibre-reinforced plastics, positrons of appropriate, well-defined energies are needed. The only practical means for producing monoenergetic positron beams of various energies is to analyse, magnetically, positrons emitted from appropriate radioactive sources. Such a technique was used by Singh et al. [40] for a preliminary investigation of moisture distribution in polymers.

1.3.7. Determination of composite stresses and fibre content with X-ray diffraction

It is difficult to determine stresses in a polymeric material directly by use of X-ray diffraction. This is due to the poor diffractive properties of such materials. It has therefore been proposed to use metallic filler particles embedded in the polymeric resin [41], [42]. By determining the strains in the filler particles, the stresses in these particles can be determined assuming linear elastic conditions, and these stresses can then be related to the resin stresses. X-ray diffraction was used in ref. [43] to determine the fibre stresses directly in an aramid-fibre composite.

This method can be used to determine applied as well as residual stresses in a composite. In ref. [44] it is demonstrated that X-ray diffraction can be used to determine the fibre content in fibre-reinforced composites. It was observed that for unidirectional glass-fibre-reinforced composites, the peak height of the relative intensity of the diffracted X-rays depended linearly on the fibre volume fraction. However, the slope of this relation was negative, implying that the peak height decreases when the fibre volume increases. This suggests that absorption phenomena are predominant, i.e. the greater the amount of glass fibres in the composite the more will be the absorption of X-rays, giving a lower peak height of the intensity plot. When carbon-fibre-reinforced composites were tested, an X-ray intensity plot with a higher peak height was obtained. This agrees with the fact that the mass absorption coefficient for carbon is lower than for glass. Also, for unidirectional carbon-fibre-reinforced composites, a linear relation between peak height and fibre volume fraction was obtained.

REFERENCES

1. Aman, J. K., Corney, G. M., McBride, D. and Turner, R. E., Fundamentals of radiography. Back to basics, *Materials Evaluation*, 1978, **36**(4), 24–32.
2. *Ibid.*, 1978, **36**(6), 24–30.
3. *Ibid.*, 1978, **36**(8), 22–32.
4. *Ibid.*, 1978, **36**(9), 19–26.
5. *Radiography in Modern Industry*, Eastman Kodak Co, Rochester, NY, 4th edition, 1980, 166 pp.
6. Domanus, J. C. and Lilholt, H., Non-destructive control of carbon fibre reinforced composites by soft X-ray radiography, *Proc. 2nd Int. Conf. on Composite Materials*, 1978, 1072–1092.
7. Blosser, E. G., McGovern, S. A. and Dhonan, O. E., *S-3A graphite/epoxy spoiler development program, Volume II*, LTV Aerospace Corp, Dallas, Texas, Vought Systems Div.; Final Technical Report No. 2-53443/4R-3172, Vol. 2, Contract No. N62269-73-C-0610. July 1975, 61 pp; Report No. NADC-75141-30, AD A031 066.
8. Hagemaier, D. J. and Fassbender, R. H., Nondestructive testing of advanced composites, *Materials Evaluation*, 1979, **37**, 43–49.
9. Prakash, R., Non-destructive testing of composites, *Composites*, 1980, **11**, 217–224.
10. Matzkanin, G. A., *Nondestructive Evaluation of Fiber Reinforced Composites*, Vol. 1 (Part 1. Radiography and Ultrasonics), NTIAC-82-1, Southwest Research Center, San Antonio, Texas, March 1982, 62 pp; AD A112 568.
11. Scott, I. G. and Scala, C. M., A review of non-destructive testing of composite materials, *NDT International*, 1982, **15**(2), 75–86.
12. Martin, B. G., An analysis of radiographic techniques for measuring resin content in graphite fiber reinforced epoxy resin composites, *Materials Evaluation*, 1977, **35**(9), 65–68.
13. Förli, D. and Torp, S., NDT of glass fiber reinforced plastics (GRP), *Eighth World Conference on NDT*, Cannes, France, 1976, Paper 4B2.
14. Crane, R. L., Chang, F. H. and Allinikov, S., The use of radiographically opaque fibers to aid the inspection of composites, *Materials Evaluation*, 1978, **36**, 69–71.
15. American Society for Testing and Materials, *ASTM STP 723, Fatigue of Fibrous Composite Materials*, 1981.
16. Blom, A. F., *Fatigue of Fibrous Composites*, Aeronautical Research Institute of Sweden, 1983, 47 pp, Report No. FFA TN 1983–30.
17. Sendeckyj, G. P., The effect of tetrabromoethane-enhanced X-ray inspection of fatigue life of resin–matrix composites, *Composites Technology Review*, 1980, **2**(1), 9–10.
18. Ratwani, M. M., Influence of penetrants used in X-ray radiography on

compression fatigue life of graphite/epoxy laminates, *Composites Technology Review*, 1980, **2**(2), 10–12.
19. Rummel, W. D., Tedrow, T. and Brinkerhoff, H. D., *Enhanced X-ray stereoscopic NDE of composite materials*, Martin Marietta Corporation, Final Report, AFWAL-TR-80-3053, Contract No. 33615-79-C3220, Air Force Wright Aeronautical Laboratories, 1980, 180 pp; AD A111 303.
20. Mohlin, T., Carlsson, L. and Blom, A. F., An X-ray radiography study of delamination growth in notched carbon/epoxy laminates, *Proc. Conf. Testing, Evaluation and Quality Control of Composites*, Butterworth's, Guildford, England, 1983, pp. 85–92.
21. Mohlin, T., Blom, A. F., Carlsson, L. and Gustavsson, A. I., Delamination growth in notched graphite/epoxy laminates under compression fatigue loading, *ASTM STP 876, Delamination and Debonding of Materials*, W. S. Johnson (ed.), 1985, pp. 168–188.
22. Blom, A. F., Fracture mechanics analysis and prediction of delamination growth in composite structures, *Proc. Fatigue 84*, G. J. Beevers (ed.), EMAS, Warley, England, 1984, pp. 1881–1891.
23. Chang, F. H., Gordon, D. E., Rodini, B. T. and McDaniel, R. H., Real-time characterization of damage growth in graphite/epoxy laminates, *J. Composite Materials*, 1976, **10**, 182–192.
24. Maddux, G. E. and Sendeckyj, G. P., Holographic techniques for defect detection in composite materials, *ASTM STP 696, Nondestructive Evaluation and Flaw Criticality for Composite Materials*, R. B. Pipes, (ed.) 1979, pp. 26–44.
25. Sendeckyj, G. P., Maddux, G. E. and Porter, E., Damage documentation in composites by stereo radiography, *ASTM STP 775, Damage in Composite Materials*, K. L. Reifsnider, (ed.) 1982, pp. 16–26.
26. Crossman, F. W. and Wang, A. S. D., The dependence of transverse cracking and delamination on ply thickness in graphite/epoxy laminates, *ibid.*, 1982, pp. 118–139.
27. Ramkumar, R. L., Compression fatigue behavior of composites in the presence of delaminations, *ibid.*, 1982, pp. 184–210.
28. Ratwani, M. M. and Kan, H. P., Effect of stacking sequence on damage propagation and failure modes in composite laminates, *ibid.*, 1982, pp. 211–228.
29. Badaliance, R. and Dill, H. D., Damage mechanism and life prediction of graphite/epoxy composites, *ibid.*, 1982, pp. 229–242.
30. Black, N. F. and Stinchcomb, W. W., Compression fatigue damage in thick, notched graphite/epoxy laminates, *ASTM STP 813, Long-Term Behavior of Composites*, T. K. O'Brien (ed.) 1983, pp. 95–115.
31. Bailey, C. D., Freeman, S. M. and Hamilton, J. M., Detection and evaluation of impact damage in graphite/epoxy composites, *Proc. 9th SAMPE Tech. Conf.*, Atlanta, Georgia, USA, 1977, pp. 491–503.
32. Sigety, P., Fatigue tolerance of carbon epoxy composites for plain, drilled and impacted specimens, Proc. 3rd Risø Int. Symp. on Metallurgy and Materials Science, H. Lilholt and R. Talreja (eds.), Risø National Laboratory, Roskilde, Denmark, 1982, pp. 311–317.

33. Peters, P. W. M., Creep and creep damage around holes in 0/90 and 0/±45/0 graphite/epoxy laminates up to 110 °C, *ibid.*, 1982, pp. 271–278.
34. More, N., Basse-Cathalinat, B., Drouillard, J. and Ducassou, D., Application of novel techniques of medical imaging to the non-destructive analysis of carbon–carbon composite materials, Proc. 4th Int. Conf., SAMPE European Chapter, Bordeaux Lac, October 1983, pp. 169–177.
35. John, J., Neutron radiography for nondestructive testing, *Sagamore Conference Proc: Nondestructive Evaluation of Materials*, Plenum Press, 1979, pp. 151–182.
36. Dance, W.E. and Middlebrook, J. B., Neutron radiographic nondestructive inspection for bonded composite structures, *ASTM STP 696, Nondestructive Evaluation and Flaw Criticality for Composite Materials*, R. B. Pipes, (ed.) 1979, pp. 57–71.
37. Dance, W. E., *N-ray inspection of aircraft structures using mobile sources: a compendium of radiographic results*, Vought Corporation, Final Report, ATC-B-92200-8CR-137, Contract No. N68335-77-C-0555, Naval Air Systems Command, April 1979, 50 pp.; NAEC-92-116. AD A068 316.
38. Singh, J. J., Holt, W. H., Mock, Jr, W. and Buckingham, R. D., Positron annihilation studies of moisture in graphite-reinforced composites, *Proc. American Ceramic Society Conf. on Composites and Advanced Materials*, Vol. 1, No. 7–8(A), 1978, pp. 473–480.
39. Singh, J. J., Holt, W. H. and Mock, Jr, W., *Moisture determination in composite materials using positron lifetime technique*, NASA Technical Paper 1681, 1980, 20 pp.
40. Singh, J. J., Holt, W. H. and Mock, Jr, W., *Positron annihilation spectroscopy with magnetically analyzed beams*, NASA Technical Memorandum 84535, 1982, 19 pp.
41. Predecki, P. and Barrett, C. S., Stress measurement in graphite/epoxy composites by X-ray diffraction from fillers, *J. Composite Materials*, **13**, January 1979, 61–71.
42. Predecki, P. and Barrett, C. S., *Detection of matrix stresses in graphite/epoxy composites by X-ray diffraction from crystalline fillers*, Final Report 3/1/78-2128181, University of Denver Research Institute, December 1981, 34 pp., AFOSR-TR-82-0174; AD A112 288.
43. Larsson, M., Schill, I. and Söderqvist, R., *Non destructive measurement of fibre stress in aramide fibre laminates by an X-ray diffraction method*, FOA 2 Rapport, C 20242-F2(F3), Försvarets Forskningsanstalt, Stockholm, Sweden, May 1978, 15 pp, N79-16928.
44. Prakash, R., Fibre volume fraction measurements in composites by X-ray diffractometer, *Composites*, July 1981, **12**(3), 193–194.

Chapter 2

Acoustic Emission

M. ARRINGTON

Speedtronics Ltd, Huntingdon, UK

2.1. INTRODUCTION

The early work on acoustic emission (AE) in the composite field concentrated on glass-fibre-reinforced plastics (GRP), but more recently the method has been extended to include carbon fibre, Kevlar and hybrid composites, as well as natural composites such as wood and bone. Most of the composites investigated have had polymeric matrices, although there has been some work carried out on metal fibre/metal matrix and carbon/carbon materials. In addition there has also been some associated work on surface evaluation and adhesion in lap joints. Most of the laboratory work involving AE monitoring of composites has been targeted at understanding, differentiating and following the various failure processes involved in composite deformation, degradation and damage. In addition there have also been some investigations into the effects of corrosion and stress corrosion. On the industrial side, AE is becoming an accepted way of testing GRP storage tanks, piping and other components, such as 'bucket trucks', whilst the associated 'stress wave factor' method is being used to detect impact damage and to monitor the curing process in composite fabrication.

This chapter is divided into two major sections: the first will provide a general survey of acoustic emission, covering key areas of the technology; whilst the second section will concentrate on the application of AE to composites.

2.2. SURVEY OF ACOUSTIC EMISSION

Acoustic emission can be likened to the materials scientists' stethoscope, allowing us to 'listen' to the low level sonic or ultrasonic signals generated by materials deformation, degradation or damage. The method can be used in several ways, as an investigative technique, to assist in understanding the mechanical behaviour of materials, as an NDT technique for assessing the structural integrity of structures and components, or as a quality control method.

Historically, a few high energy AE processes have been observed directly, such as 'tin cry' (twinning), 'clinking' in castings (martensitic transformations), 'dunting' in ceramics (thermal shock) and creaking of pit props. Many more materials are AE active, but at lower energies. In Drouillard's bibliography [1] the list of materials investigated runs from aluminium to zirconium.

In order to detect and process these low level events, it is necessary to use a sensitive detector (sensor) in conjunction with high gain analogue electronics. The technology, in its complete sense of involving both the detection, processing and analysis of the AE activity for the systematic investigation of materials, started early in the 1950s. Although there were some earlier, isolated, experiments, Josef Kaiser, who worked at the technical high school in Munich, is generally recognised as the father of modern AE. Over the past 30 years, AE has developed rapidly from a laboratory technique to an industrial NDT method. Much of the early work concentrated on metals, but the more recent work has been much wider, covering composites, ceramics, plastics and process monitoring.

Before describing the specific applications of AE to composites and composite structures, it is worthwhile reviewing the main features of the technique, in order to give the reader an appreciation of the various factors involved in AE testing.

If we examine a simple test, such as K_{1C} (fracture toughness) test, we can get some idea of the wide spectrum of disciplines involved in AE work. When we apply a load to the specimen it starts to deform elastically; associated with this deformation is the storage of elastic strain energy. When the starter crack extends, some of the stored energy is released. Part of this released energy is absorbed by the surface energy required to produce the new crack surface, some by plastic zone growth, and some produces the AE activity (stress waves). These stress waves travel from the source to the sensor, which will receive not only the direct waves but also

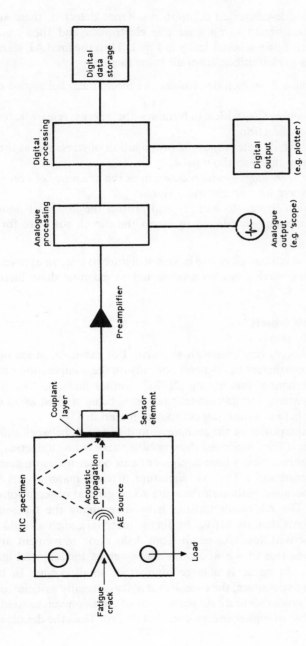

FIG. 1. Schematic illustration of acoustic emission generation, detection and analysis.

reflected and mode-converted components. Once detected, these signals have to be conditioned using analogue electronics and then analysed digitally. This is shown schematically in Fig. 1. Thus the final AE signature will depend on contributions from all these factors:

(1) *Material science*—which determines the mechanical behaviour of the test piece;
(2) *Fracture mechanics*—which determines the energy release rates and the partition function;
(3) *Acoustics*—which determines the propagation of stress waves through the specimen and into the sensor;
(4) *Transducer technology*—which determines the efficiency of converting the stress wave into an electrical signal;
(5) *Analogue processing*—to amplify and protect the low level signal;
(6) *Digital processing*—hardware to assess the signal, software for data analysis.

If, instead of a small test piece, we have a structure to test, an appreciation of structural engineering is also needed. Let us examine these factors in more depth.

2.2.1. Materials aspects
2.2.1.1. *Introduction*
Acoustic emission is very materials sensitive. For example, in aluminium alloys, the AE characteristics depend not only on the composition but also on the state of temper and ageing [2]. In a similar fashion, AE, in conjunction with pattern recognition techniques, is being studied as an aid to differentiating between differing composite materials [3].

This shows the power of the technique to distinguish between different microstructures. (The associated drawback is there is no 'universal' calibration as in ultrasonics, where equipment can be set up using standard reflectors as references.) The AE signature of each material has to be established separately, although there are some general observations that can be made. The AE characteristics help in resolving the types of deformation process that are active. In ductile materials, such as mild steel, the emission activity tends to result from dislocation movement around yield. This gives rise to a smooth emission versus load plot (Fig. 2a), indicating that the cause is a large number of small events. In brittle materials, such as ceramics, the emission activity is usually associated with cracking. This gives rise to a 'staircase' plot of emission versus load with a smaller number of higher energy events (Fig. 2b). Thus the details of the

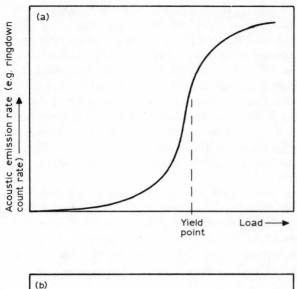

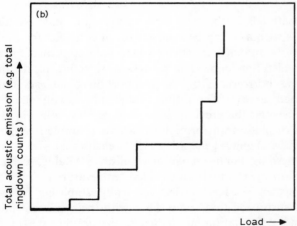

FIG. 2. Schematic illustration of acoustic emission activity plots. (a) From ductile materials; (b) from brittle materials.

AE response assist in separating and identifying different deformation mechanisms. Composites have a complex set of possible deformation and failure mechanisms and can thus be expected to have complicated AE signatures. This will be covered in more detail in Section 2.3.

2.2.1.2. *Kaiser effect/Felicity effect*

One important effect observed in AE monitoring is the 'Kaiser' effect. This is, that a repeat loading to a previously experienced load level gives rise to no additional AE activity, as illustrated in Fig. 3. This is a somewhat simplified description, although it works well for immediate reloads on materials within their elastic limit, where no additional damage is being caused on reloading. To appreciate the more general case it has to be realised that it is the local stress that is the driving force behind the AE activity. Thus if we examine a sample, like the K_{1C} specimen with a crack in it, then the local stress at the crack tip is the product of the macroscopic stress (i.e., load/area) and the microscopic stress intensity factor (i.e., crack geometry).

If the crack extends, for example by fatigue, between load interrogations, then, on reloading, the stress at the crack tip will be higher than before; this is because the stress intensity factor will have increased, thus allowing renewed AE activity. This forms the basis of an AE-based periodic inspection programme. The Kaiser effect breaks down in several circumstances:

(i) When materials lie beyond their elastic limit and time-dependent processes such as creep become active. In concrete, for example, we can see a 'supra' Kaiser effect, where holding a load for a period, during which time-dependent processes are active, results in AE activity being 'borrowed' from the next portion of the instantaneous AE versus load curve [4]. This results in a quiet period on reloading which extends beyond the previous load level, as shown in Fig. 4.
(ii) In cases of phase transformations, where a sample may go through many phase changes generating AE at each cycle.
(iii) Where there are damage repair mechanisms available, such as annealing in plant operating at an elevated temperature.
(iv) In composites. The Kaiser effect was really established for metals and it is not surprising that composites, with their different microstructures and deformation mechanisms, do not behave in the same fashion as isotropic materials. This breakdown of the Kaiser effect in composites is called the 'FELICITY' effect [5], and forms an important part of the CARP code. These factors illustrate that, as in all AE work, we have to be aware of the materials' science aspects involved.

2.2.2. Degradation mechanisms

The driving force behind all AE activity is the release of energy stored

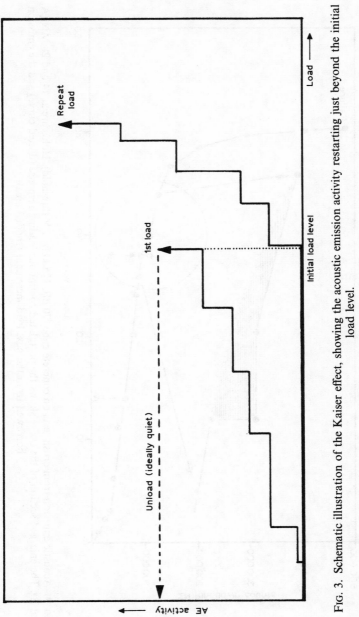

FIG. 3. Schematic illustration of the Kaiser effect, showing the acoustic emission activity restarting just beyond the initial load level.

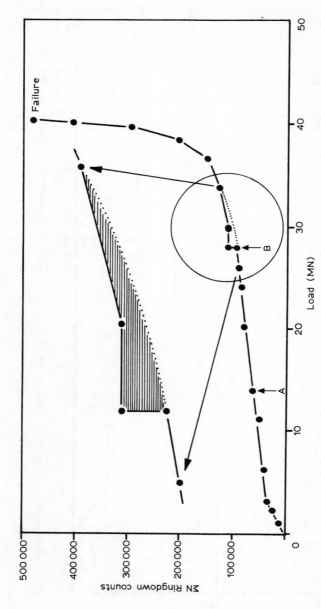

FIG. 4. Acoustic emission from crush tests on concrete cube (70 dB, 175 kHz, 1 V threshold). Holds are at 14 and 28 MN (A and B). The enlarged section of the curve shows the postulated continuous loading response (dotted curve) and the emission borrowed from the next load increment (shaded area).

within the test piece, component or structure, hence the connection with fracture mechanics. However, there are other energy sources available besides elastic strain energy. Table 1 lists some of these sources and the mechanisms that result in energy release and AE.

TABLE 1
Energy Sources and Release Mechanisms

Energy stored as	Energy released by
Elastic strain energy	Deformation and crack growth
Chemical potential	Corrosion
Crystalline free energy	Phase transformations

Another series of mechanisms can also give rise to signals in the AE spectrum. These tend to be process oriented such as leakage, machining or machinery noise. These mechanisms offer the potential to use AE for monitoring machinery health, condition monitoring and, in the longer term, process control.

In most of these various mechanisms only a fraction of the total energy released is available to generate AE (stress waves). Other energy sinks are also in contention for the released energy and some form of partition function will determine the distribution of the available energy between

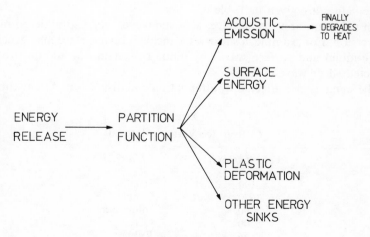

FIG. 5. Partition function.

the various sinks (Fig. 5). The stress wave so generated is a transient affair, in that it rapidly degrades to heat energy due to attenuation.

2.2.3. Acoustic aspects

Once the stress field around the crack tip rearranges, it produces a stress pulse that travels through the specimen. This is where the acoustics come into AE. Acoustic propagation is quite complex, involving:

(1) different wave types;
(2) mode conversion and reflection;
(3) attenuation;
(4) dispersion.

In an infinite unbounded solid medium only two wave types are observed, longitudinal and shear waves. The propagation velocities of these waves are determined by the mechanical properties of the solid. Once boundary conditions are imposed by the introduction of free surfaces, families of more complicated wave types become allowed, such as surface waves, plate waves and rod waves. All these have their own individual characteristics, such as propagation velocity and attenuation, as a result of their different mechanical and microstructural properties. For example, one of the simplest waves from the propagation velocity point of view is the longitudinal rod mode. This has a velocity

$$C(\text{rod}) = \text{square root (Young's modulus/density)}$$

Some values are given in Table 2.

On reflection at a free surface, acoustic waves can also undergo mode conversion. For example, a shear wave incident on a surface may generate longitudinal and surface waves by mode conversion in addition to the reflected shear wave (Fig. 6).

The signal is also affected by attenuation and dispersion. Attenuation is

TABLE 2
C (Rod) for Various Materials

Velocity (m/s)	Material
5190	Steel
3670	Copper
5090	Aluminium
5300	Glass
46	Rubber

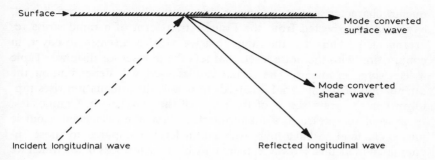

FIG. 6. Mode conversion at a free surface.

frequency-dependent, with the higher frequency components suffering more losses. Whilst some modes are dispersive (i.e. different frequencies propagate at different velocities), the effect of these two factors is to degrade the signal, slowing down the rise time, stretching the pulse length and depressing the peak amplitude.

Acoustic propagation in composites is itself made more complicated by the anisotropic nature of the materials involved. Thus the longitudinal and shear waves, which are non-dispersive in isotropic materials, become directionally dependent in composites. For more information on stress wave propagation in general see *Stress Waves in Solids* by Kolsky [6], whilst the effect of composite composition (volume fraction/porosity) is discussed by Reynolds and Wilkinson [7] and anisotropic propagation by Glennie and Summerscales [8].

2.2.4. Sensor aspects

The sensors most commonly used in AE are resonant piezoelectric devices, although there have recently been some experiments on using PVDF (a piezoelectric polymer) as a sensor [9]. In addition there has also been some work on the use of optical techniques, such as interferometry, to detect AE activity in composites [10].

Most AE work is carried out in the low ultrasonic range 100–500 kHz, although for special circumstances other frequencies may be used; for example, low frequency sensors to cope with high attenuation situations, or high frequency sensors to escape from background noise problems. The AE spectrum is shown in Fig. 7.

The choice of operating frequency depends on several factors: the ambient background noise level, the material's attenuation and the frequency characteristics of the source. In general the source is broadband, as would be expected from the Fourier transform of a rapid stress rearrangement. Thus it is the range that we wish each sensor to cover, in conjunction with the noise level, that sets the intersensor distance. Table 3 lists some examples. The actual ranges used will depend upon the attenuation of the material involved. In metals the attenuation rises rapidly when the grain size is of the order of the wavelength. Composites, because of the presence of many interfaces and the associated acoustic mismatch, tend to have high attenuation levels compared to those encountered in metals. However, there may be some low loss paths which we can make use of. For example, in pressure vessels we can immerse a sensor inside the vessel and improve the coverage [11].

There are two main types of sensor, broadband and resonant. The broadband devices can be used for spectrum analysis, although care must be taken to ignore specimen resonances. Most AE work is however carried out using resonant sensors. This has the dual advantage of greater sensitivity combined with reduced background noise, due to narrow band operation. (Noise in this context consists of both electronic noise and the ambient acoustic background in the working environment.)

Sensors are often used in a differential form in conjunction with a differential amplifier to reject electromagnetic interference (EMI). One other important aspect of using sensors is the acoustic coupling between sensor and specimen. For stress waves normally incident at a steel/air interface 99% of the energy is reflected back into the steel. If the air is replaced by water approximately 90% of the energy is transmitted into the water layer. Thus we use acoustic couplants such as resins, greases and oils to assist the stress wave to propagate out of the specimen into the sensor.

Sensor calibration is a complex task, and various methods have been suggested. These include ultrasonic-style reciprocity techniques [12], the use of helium gas jets to generate a stable broadband signal [13], as well as impulse methods based on sparks, glass capillary failure and pencil lead breakage [14-16].

Most sensors are small cylinders approximately 1 inch diameter by 1 inch long, although miniature sensors are available for use on small samples. Since the sensors are manufactured from PZT they cannot generally be used at high temperatures. PZT5A for example has a Curie temperature of 365 °C; this is the temperature at which the material has completely depoled. Such material should not be used at over 200 °C for any extended period. If it is necessary to carry out high temperature tests,

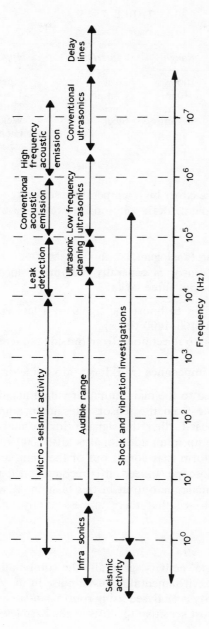

FIG. 7. Acoustic emission and the sonic/ultrasonic spectrum.

TABLE 3
Intersensor Distances

Frequency	Range	Material	Application
30 kHz	100 m	Steel	Pipelines
75 kHz	10 m	Composites	Tanks
175 kHz	10 m	Steel	Tanks
375 kHz	1 m	Steel	Welds
750 kHz			High noise situations

then it is possible to use either high temperature sensors or waveguides to provide a thermal gradient. Both solutions, however, result in some loss of sensitivity.

2.2.5. Signal processing (analogue)
The signal from the sensor is generally fed into a local preamplifier (preamp) which carries out three tasks:

(1) to provide some gain to boost the signal to a less vulnerable level, typically 40–60 dB (100–1000 times);
(2) to provide filtration to reject noise from outside the sensor's operating range;
(3) to match the high-impedance transducer into a line driving situation.

The signal is then fed to the main amplifier which may be up to a few hundred metres remote from the sensor. This main amplifier provides additional gain to boost further the signal prior to analysis. The main amplifier may provide up to an additional 60 dB (1000 ×).

The analogue waveform that comes out of the main amplifier can be displayed on an oscilloscope. Indeed, it is recommended that a scope is used to monitor the analogue output in this fashion. If we examine the scope display, we will see that there are essentially two AE regimes (Fig. 8).

2.2.5.1. Continuous activity
This is the 'steady state' activity, generally the combination of the electronic noise plus the environmental acoustic noise in the AE bandwidth. Such continuous activity is analysed using root-mean-square (RMS) techniques. This will be most sensitive to shifts in the base level of the signal.

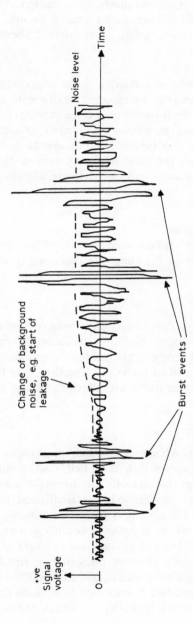

FIG. 8. Illustration of continuous/burst activity. The continuous activity is the sum of the electronic noise and the ambient acoustic noise. If the background level changes, due to leakage for example, the apparent noise level will vary. The use of an automatic threshold that maintains a constant deadband above the noise level will assist in resolving the burst events from the background.

Thus RMS is used to detect leaks and plastic deformation. At high event rates, such as those observed in composites around failure, RMS techniques can still follow the activity where some other methods fail.

2.2.5.2. *Burst activity*
Superimposed over the continuous activity is the burst activity. These bursts correspond to the discrete micromechanical events that are occurring within the material. Such bursts tend to be random in timing and size, and it is usual to separate them from the continuous activity. This is usually achieved by the use of a threshold; only events that exceed the threshold are processed. Since the noise level may vary, some equipment uses a 'floating' threshold in order to improve the resolution of the burst events.

2.2.6. Signal processing (digital)
Various features of the waveform are used to make digital measurements upon the burst waveform (Fig. 9). These are described in the following paragraphs.

2.2.6.1. *Event counting*
This is the simplest method of signal analysis, scoring one count per event. Electronically this is often achieved digitally: the chain of threshold crossing pulses is inspected for a time gap greater than a specified length; this is then used to indicate the end of the event. The time may be chosen so that fine structure and reverberations are allocated to the same source-event.

2.2.6.2. *Ringdown counting*
This is a very commonly used AE signal assessment technique. As an event hits the sensor it makes the sensor ring like a bell. Each oscillation of the resultant electrical waveform that exceeds the threshold scores one ringdown count. As the sensor is damped, the oscillation decays in an approximately exponential fashion; it is from this 'ringing down' that ringdown counts get their name. Ringdown counts give an idea of the energy content in the event, but not an accurate measure of energy.

It is sometimes useful to look at count rates such as ringdown counts per second, especially in high noise situations, as this allows an average background level to be established from which excursions caused by AE can more easily be distinguished. In a similar fashion, the counts plotted

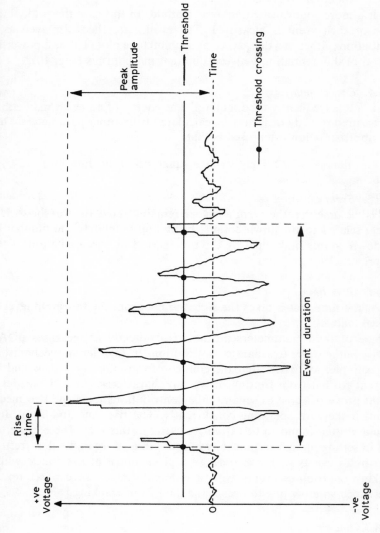

FIG. 9. Burst waveform analysis methods.

per increment of such parameters as stress or strain can also be more illuminating than the simpler total count versus time or load plots.

2.2.6.3. *Amplitude analysis*
This is a more sophisticated analysis method. In this case the peak amplitude of each event is measured. The results are then displayed as a distribution; either as a density, or as a cumulative plot. This is a powerful method of differentiating between different mechanisms (Fig. 10).

2.2.6.4. *Energy analysis*
This is a less frequently used method. The energy of an event may either be measured directly or it may be calculated from other parameters. One such approximation often used is that

$$\text{Energy} = 1/2 \text{ (Peak voltage squared} \times \text{event duration)}$$

2.2.6.5. *Event duration*
This is the time that the event envelope remains above the threshold. It is closely related to ringdown counts, being approximately the number of ringdown counts multiplied by the time period of one oscillation of the sensor.

2.2.6.6. *Rise time*
This is the time that it takes the signal to rise from the threshold level to its peak value.

Some of these parameters are not used directly as measures of AE activity but in order to separate contributions from different mechanisms. For example, we expect that the emission from crack growth would be different to that from fretting. If we can characterise these signatures it should prove possible to separate the contributions from the two mechanisms if they are operating concurrently. The rise time and event duration are often found to be extremely useful in this role. These are often used in setting up an acceptance 'window'; thus, if we know that fretting contributes events with rise times in a given range of rise times, whilst cracking contributes events with a different range, we can set up an acceptance window just to accept the data from cracking [17].

2.2.6.7. *Source location*
So far all this signal analysis has been oriented at single channel equipment, but one of the strengths of AE is its ability to locate the sources

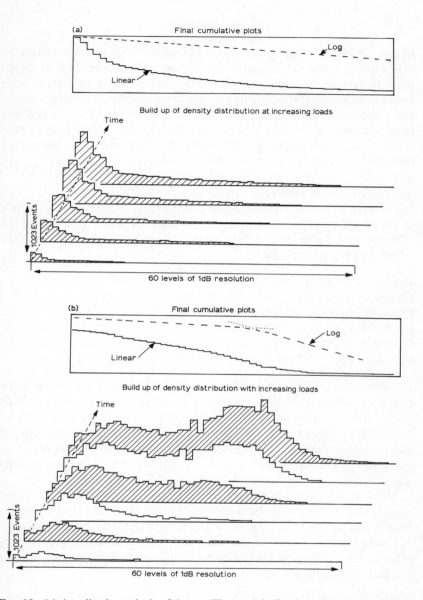

FIG. 10. (a) Amplitude analysis of tin cry. The straight line logarithmic cumulative plot is indicative of a single AE source. (b) Amplitude distribution analysis of composite testing. The twin gradients in the logarithmic cumulative plot is indicative of the two AE sources, the low amplitude source at low loads and the higher amplitude source operating at higher loads.

of the AE events using multichannel equipment. This involves having an array of sensors and measuring the arrival times at the various sensor sites. In fact we measure the 'delta t's—the difference in arrival times between pairs of sensors. Using triangulation techniques, the location of the source can then be calculated. The equipment used for this task will have a computer to carry out the calculations. It has been found that 'clustering' of a number of AE events in a small area is a very good indication of a localised anomaly in a structure.

2.2.6.8. *Choice of equipment*
AE is often a contest between signal and noise; in order to give ourselves the best chance of success we try to enhance the signal and reduce the effect of the noise at each stage in the aquisition and analysis of data. The techniques we use include:

(1) choice of sensor frequency and bandwidth;
(2) use of differential sensors and preamp;
(3) use of floating threshold;
(4) use of acceptance windows;
(5) use of location and clustering;
(6) use of any other information available.

For example, in fatigue testing, it is often found that the test-rig noise is at a maximum when the crosshead is in motion and that, at peak load, there is a valley in the noise. This is just where we expect the AE associated with fatigue damage to occur. Thus, by enabling the AE system only to accept data around peak load, we can, in effect, improve the signal to noise ratio. The equipment used to implement this is called a voltage controlled gate and is triggered from the load cell signal.

There is one other more specialised signal processing and analysis technique that is used for fundamental studies. This comprises extremely broadband equipment used in conjunction with specially designed samples to detect the initial part of the AE stress wave before it is coloured by reflections or specimen resonances. Such equipment also offers capabilities for good spectrum analysis.

2.2.7. Data recording
The data recording facilities of a system tend to reflect its level of sophistication. Data recording falls into several types, as described below.

2.2.7.1. Analogue (*real time*)
Using a modified videotape, it is possible to record the analogue output from the main amplifier. This approach does suffer some drawbacks, since such units tend to have limited dynamic range; nevertheless, they can still be very useful.

2.2.7.2. Digital (*real time*)
Using a transient recorder or a fast A-to-D converter, the signal can be captured digitally. The problem here is the rapid sample rate and wide dynamic range involved (typically 5 MHz sampling rate and > 60 dB range), which lead to a large amount of raw data with a corresponding mass storage problem for all but the shortest test.

2.2.7.3. Analogue (*processed data*)
Most of the simple single channel units will provide analogue outputs to drive chart recorders and X-Y plotters. These can be purely analogue outputs, such as RMS, or they may be digital data such as ringdown counts or amplitude distributions that have gone through an internal D-to-A conversion.

2.2.7.4. Digital (*processed data*)
Most of the computer-based equipment will record onto disk the AE characteristics of each individual event. This allows playback of a test, with the option of varying the test set-up or parameters on post-test processing.

The more advanced rack systems will allow access to the accumulated digital data, stored in the hardware counters, via a computer interface. This allows for external data logging. There are two main differences between these two approaches:

Software. With computer-based systems the software must be realtime, whilst a rack system in the data logging mode does not require realtime programs.

Event characterisation. In a computer-based system the complete AE description is available for each event on an event-by-event basis. In the simpler systems such correlation may not be possible as the various stores are only accessed relatively slowly. Thus only average values may be available.

Thus, if we want to store AE data, we must ensure that there is sufficient storage available. This in turn means that the more processing that can be accomplished in real time prior to recording the better, since this both reduces the storage requirement and adds to the ease of reprocessing the data that we do choose to store. If we fill a 15 Mbyte hard disk with 500 kbyte of useful data and 14.5 Mbyte of noise, this is going to act as a significant deterrent to reprocessing the good data. Our experience in the offshore environment was that data validation, through acceptance windows, was a powerful way of concentrating on useful data. In a two week period there could be up to 1 million system interrupts (one or more sensors being triggered). Of these, only 1000–2000 events would be accepted as of interest; that is, coming from the areas of interest (the weld lines) and having the specified emission characteristics. (These were set up as the result of some laboratory trials monitoring crack propagation.) This is likely to be an important consideration when testing composites as they can give rise to high data rates, especially at failure and beyond.

2.2.8. Methodology behind the use of AE

AE has been used at a variety of levels of sophistication, depending upon the inspection facility required. These tend to reflect the number of channels involved and the test duration. Some examples are as follows.

2.2.8.1. *Laboratory/materials testing*

This can often be accomplished using either a single channel or a small multichannel system. It is often an advantage at this stage to have a computer-based system to allow for data recording and reprocessing. However, once a test procedure has been established in this way, it is often possible to implement such a test on a simpler (and hence less expensive) rack system.

2.2.8.2. *Quality assurance (QA) testing*

This can often be the result of a previous phase of materials testing. Such QA testing is to ensure that a material is to specification or that a component is performing within its mechanical specification. Examples of this are:

(1) the inspection of incoming billets by a Swiss institute prior to releasing the material for use in watch springs;
(2) the testing of glass bottles to ensure that there are no cracks under the closure (an area inaccessible to the more conventional optical techniques).

2.2.8.3. *Proof testing*

Many of the larger AE tests are concerned with monitoring proof tests on metal and composite vessels. This makes use of the large multichannel computer-based location systems. The objectives of such tests are to locate problem areas and to assess the quality of the vessel. The basis of proof testing is that, by surviving a pressurisation beyond the working pressure, there is a margin of safety between the maximum possible crack that will allow the structure to survive the proof load and the critical crack size at working pressure.

AE gives us the extra ability to assess the damage caused by the proof test itself, and to grade the surviving vessels. This is an area where we are starting to see codes of practice becoming established. This is an important step in gaining acceptance in the general engineering community. The CARP code for composite inspection will be covered later in Section 2.3.4. Such proof testing can be on a one-off acceptance basis or as part of an ongoing periodic inspection programme.

2.2.8.4. *Periodic inspection*
Using the Kaiser effect approach, it is possible to establish a periodic inspection programme to detect structural degradation. This aims at identifying plant that has degraded either by fatigue or corrosion since the previous test.

2.2.8.5. *Continuous monitoring (structural and process)*
AE offers a powerful continuous monitoring option. Although this is also the most demanding situation for the instrumentation, AE has been used in this fashion by British Steel to monitor the headers on the Redcar coke ovens. With computer systems such continuous monitoring may involve the use of uninterruptible power supplies to reduce the incidence and effects of power cuts and mains fluctuations. In general we can anticipate that such on-line monitoring will tend to concentrate on trouble spots (either known or suspected). The use of a single data channel, in conjunction with an appropriate array of guard sensors, can provide an economic solution to the provision of such a facility.

It is to be expected that such monitoring would often be called for by the petrochemical industry, where hazardous environments are often encountered. AE can now be used safely in such environments, due to the recent availability of intrinsically safe front ends (sensor/preamp/cable combinations approved by BASEEFA and CENELEC).

2.2.9. Comparisons with other NDT techniques
The unique features of AE when compared to the other NDT techniques are:

(1) The ability to detect and locate defects from a few static sensor

locations. Because of the relatively low ultrasonic attenuation in the typical AE frequency range (hundreds of kHz), it is possible to detect emissions that originate perhaps several metres away from the sensor. This means that we can achieve 100% coverage of a structure from a relatively few sensor sites. This in turn means that an AE test can be quickly set up on site.

(2) The link between AE activity and stress intensity. This is best shown by the high strength materials for which linear elastic fracture mechanics works well; but the implications of such a link are important. When using the other NDT methods the crack size may be estimated if the orientation of the crack is favourable. This is then put into context, as far as structural integrity is concerned, by the use of stress analysis, to determine whether such a defect is dangerous or not. Such stress analysis is not always easy, especially for complex composite structures where the loads involved are not always well established. One other important factor here is that it is the locations with short critical crack lengths (i.e. highly stressed regions), often around cutouts or profile changes, which are the most difficult to inspect using for example ultrasonics. These are just the areas which tend to favour AE monitoring.

2.2.10. Related techniques

A related technique is the 'stress wave factor' method. This involves the injection of an ultrasonic signal into a specimen and its subsequent detection and analysis using AE methods. In essence this is making an attenuation measurement on the material. Such attenuation measurements have been used to follow the deterioration of metals during fatigue [18]. The stress wave factor technique, which generally uses the lower AE frequency range (typically 100–400 kHz), extends this measurement approach to the more highly attenuating composite materials [19].

This concludes what must be a somewhat limited survey of AE. There are many papers on the various aspects of the method, but as yet there is no single source that provides a complete coverage. Thus perhaps the best key to identifying useful papers for further information is Drouillard's bibliography [1].

2.3. ACOUSTIC EMISSION FROM COMPOSITES

2.3.1. Introduction

In this section we will examine the application of AE techniques to composite testing and inspection. This will be divided into three main topics: materials evaluation, component and structural testing and process/production monitoring.

A recent bibliography [20] lists nearly 800 references on AE monitoring of composites. This shows the topic has become well established. Further confirmation of this comes from the Society of the Plastics Industry Inc., who have become involved with AE in two ways: firstly, through their support of the CARP codes for the application of AE to composite materials [21, 22] (these are issued by the Committee on Acoustic Emission from Reinforced Plastics, a working group of the Corrosion Resistant Structures Committee of the Reinforced Plastics/Composites Institute of the SPI); secondly, through their sponsorship of conferences specifically on AE from composites [23, 24].

To date most of the AE work on composites has concentrated on fibre-reinforced thermosetting plastics, although there have also been some investigations into thermosoftening plastic and natural-rubber-based materials. In addition, there has been some work done on metal matrix composites, particulate reinforced materials, and the separate constituents involved (matrix, fibre and interface). These other topics are, however, outside the scope of this review.

A wide range of mechanical tests on composites has been monitored using AE. Indeed, this is a necessary condition in understanding the AE signature of these materials, because of the differing deformation and failure processes involved in tension, bending, torsion and shear. The early work concentrated on simple uniaxial material tested in tension along the fibre direction. But it soon became apparent that the AE signature was influenced by the mode of testing, with the resultant extension in the range of mechanical testing and associated AE monitoring that has been undertaken. This factor has practical implications for the users of AE: the material, component or structure must be mechanically interrogated in an appropriate fashion, in order to provide an adequate simulation of the stress system encountered in use.

When performing mechanical tests, it is generally found that the screw-driven machines generate significantly less background noise than the servo-controlled hydraulic fatigue machines which, in turn, are quieter than the higher frequency resonance machines. There are measures that can be taken to reduce the generation and influence of machine and grip noise. These fall into three groups, as detailed below.

2.3.1.1. *Electronic means of noise rejection*
This can involve choice of sensor operating frequency, the use of signal

recognition techniques such as front end filtering, and the use of noise rejection techniques such as load gating and guard channels.

2.3.1.2. *Acoustic damping*
This involves the use of damping materials on the load-train to isolate the sample from the machine noise.

2.3.1.3. *Mechanical design of grips and specimen*
The grips can utilise materials of differing acoustic properties in order to achieve an acoustic mismatch which will reduce noise propagation out of the load frame and into the testpiece. Good grip design can also minimise the generation of any spurious activity caused by deformation or slippage.

Once we go to more complex situations than that of uniaxial material tested along the fibre direction, there are two additional factors to take into account. These are:

(1) *Material morphology*. Just as alloys with similar compositions but varying heat treatments have different AE characteristics, so do composites with differing fibre orientations and laminate build-up.
(2) *Orientation effects*. Varying the angle between the fibre orientation and the test axis used for loading will affect the mechanical behaviour and hence the AE signature.

Relatively little work has been carried out on the long term behaviour of the AE activity of composites under conditions of creep and environmental attack, although there have been several research programmes upon the effects of fatigue testing.

2.3.2. Materials evaluation
If we examine a fibre composite and the failure mechanisms available, we find that these are very different to those active in metals; thus it is not surprising that the AE behaviour of composites also varies from that of metals. Indeed, one of the early controversies in AE was over whether the Kaiser effect held for composites, and, if not, what was the cause of the breakdown. As I see it, the Kaiser effect is based upon experimental observation rather than being an intrinsic law. Thus observations made on one material do not necessarily hold for other materials. Nevertheless the Kaiser effect does hold for many metallic systems, and so it has perhaps unconsciously been promoted to an implicit law. Thus when composites were found not to obey the Kaiser effect, there was perhaps an overreaction. The Felicity effect is the complement to the Kaiser effect, and the associated Felicity ratio, which is a measure of extent of the failure of the

Kaiser effect, can be used to evaluate composite structures as it gives a measure of the severity of previously induced damage. The Felicity ratio is defined as

$$\frac{\text{Stress at which 'significant' emission restarts}}{\text{Previously applied maximum stress}}$$

The key factor in this is the definition of 'significant' emission, as it is on this feature that the accept/reject criteria are based. To some extent this is a matter of experience, although there are some guidelines. ASTM suggests:

(a) More than 5 bursts of AE during a 10% increase in load.
(b) More than 4% N_c counts during a 10% increase in load*.
(c) Continuation of AE at constant load.

CARP offers similar guidelines, the major variation being:

(b') More than 20 counts during a 10% increase in load.

There are several regimes of failure: the microscopic events that accrue prior to failure; the macroscopic processes involved both around failure and the post-failure mechanisms. Some examples are:

(1) *Initial (low load) microscopic damage.* Such damage will not necessarily reduce the short term strength of the material, but may make it vulnerable to subsequent degradation by fatigue, corrosion or SCC. Typical mechanisms include matrix/fibre debonding and matrix cracking.
(2) *Subsequent (medium/high load) microscopic damage.* Such damage will tend to reduce the short term strength of the material and thus act as a precursor to failure. Typical mechanisms include fibre failure and fibre pull-out.
(3) *Macroscopic damage (medium/high load) up to peak load.* Such damage will tend to be some form of cooperative deformation, such as cracking, delamination or crushing.

Finally, there will be the damage which occurs beyond peak load. However, if we are concerned with AE as a non-destructive test (NDT) method, then we are concerned with the performance up to peak load. (However, from a composite mechanics point of view, the behaviour

---

*Here N_c is defined as five times the total counts from thirteen 0.3 mm, 2H Pentel pencil lead breaks, for each of two locations in the region of the AE activity. This is in effect a sensitivity calibration to define the critical level of energy release.

beyond peak load may be of interest if we want to try to improve properties such as work of fracture, impact strength and ILSS [interlaminar shear strength].) This was one of the points made by the Sussex Group, that it was the work of fracture up to failure that determined the utility of a component.

Except for a few papers published in the 1960s, work on AE characterisation of composites did not really get under way until the early 1970s. This was some ten years behind the application of the technique to metals. As such the composite workers were perhaps spared some of the early teething troubles, but this time lag did result in some bias towards a metallurgically oriented interpretation of AE data; especially as far as the Kaiser effect was concerned. Because of the complexities of composite mechanics, the theoretical modelling of composite AE also lags behind that of the metals. An early model for AE from fibre cracking was presented by Harris et al. [25]. This was tested using a whisker-reinforced material (Al_3Ni in Al). The theoretically derived curves were smooth, whilst the experimental traces comprised a series of steps. This illustrates that the AE activity is significantly influenced by local features, such as composition, orientation and defects, which are not adequately represented by continuous macro-mechanics. A more comprehensive model has been presented by Henneke and Jones [26].

2.3.2.1. GRP

Experiments have been carried out on short term tensile tests [27], and on torsional and flexural specimens in GRP [28]. Long term investigations have included creep [29], fatigue [30] and environmental testing [31].

Early work by Sims et al. [32] at the National Physical Laboratory (NPL) demonstrated an ability to obtain useful correlations between total crack area (observed optically), mechanical hysteresis (damping) and the associated AE activity. Subsequent work by Harris et al. [33] indicated that there was a less clear-cut correlation between AE and damping, perhaps an indication of the influence of directional aspects in damage accumulation. More recent tests have tried to control the failure mechanism more strictly, for example by using the Iosipescu shear test to eliminate fibre failure. In this test the specimen fails by a combination of interfacial failure and matrix cracking [34]. Single fibre specimens have also been used to help in resolving the AE characteristics of specific mechanisms [35].

2.3.2.2. CFRP

Perhaps because of the higher technology applications, work on CFRP has tended to be more mechanistically oriented. Thus Carlsson has investigated delamination [36], Charentenay ILSS fatigue [37] (indeed, his work on short beam shear tests has been adopted as the basis of a materials QA test), Flitcroft and Adams have investigated shear cracking [38] and Williams and Lee have studied the effect of paper inclusions [39]. Wevers *et al.* have used energy distributions to set up a damage accumulation theory, comparing AE with other inspection methods (replication and X-ray radiography) [40].

2.3.2.3. *Other materials*
Other materials investigated include hybrids [41], wood [42] and Kevlar composites [43]. Although these other systems have not been studied as extensively as GRP and CFRP, the extension of AE monitoring from uniaxial materials to cross-ply and model structures has provided a useful step in the progression towards component and structural testing.

2.3.3. Source characterisation
Acoustic emission has been studied in order to separate and recognise the occurrence and presence of the various degradation mechanisms. Several techniques have been used in order to achieve this resolution. The early work tended to concentrate on simple waveform pattern recognition [44], but the more recent work uses more analytical methods.

2.3.3.1. *Amplitude distributions*
Amplitude distributions can be analysed either on the basis of the density distribution or by 'b' value analysis of the logarithmic cumulative distribution. Such an analysis has its origins in seismology and has the form:

$$N(a) = (a/a_0)^{-b}$$

where $N(a)$ is the fraction of the AE population with its peak amplitude in excess of a, a_0 is the lowest detectable amplitude and the exponent b characterises the distribution [45].

Wolitz *et al.* [46] have used b value analysis of the amplitude distribution to differentiate between matrix and fibre failure in tubular GRP specimens. A study by Eikelboom on CFRP found three separate peaks in the amplitude distribution [47]. These were assigned to matrix cracking, delamination and fibre failure. This assignment was achieved by the use of samples with differing fibre orientations, which resulted in changes in failure mode. A more detailed examination of this has been carried out by Guild *et al.* [48]. Amplitude analysis has shown significant promise as a method of source characterisation.

2.3.3.2. *Frequency analysis*

When using resonant sensors, we are only sampling a portion of the AE activity, which may, in general, extend from virtually DC to 30 MHz or more. Frequency analysis offers the possibility of extracting more information on the AE source by accessing the complete stress wave. However, the influence of specimen resonances and attenuation can quickly distort and colour the event. Early work by the Harwell group using special wide bandwidth equipment (0–30 MHz) confirms this [27], with the AE energy being confined to the 0–1 MHz region. Subsequent workers have tended to use more conventional AE equipment in conjunction with broadband AE sensors (typically 100 kHz–2 MHz); Williams and Egan [49], for example, have used a hardware spectrum analyser. Graham has adopted in his multi-parameter system a series of parallel bandpass filters to separate the frequency components in each event [50]. Henneke found that there were indications of spectral differences in the 0–300 KHz range, between specimens failing by fibre fracture and those failing by matrix cracking [51]. Thus, although the method has certainly produced some interesting results, spectrum analysis has not, as yet, offered any practical assistance in the task of composite inspection.

2.3.3.3. *Comparison of techniques*

The implication of the use of such analysis techniques for structural monitoring is that frequency analysis requires the use of broadband sensors located close to the failure site. This is an unattractive option, due to the extra number of sensors and other equipment that would be required to implement such a test. Thus amplitude analysis offers a more practical approach to structural monitoring, although for component testing some sort of frequency analysis may be appropriate. Full Fourier analysis can be a demanding task in terms of both the hardware and software involved; an intermediate option is to use a series of bandpass filters. An example of such equipment is provided by Graham's multiparameter AE system.

In well controlled laboratory tests, the consensus of opinion is that AE measurements do indeed allow us to differentiate between mechanisms. However, the extension of such laboratory findings (often based upon uniaxial data) to crossply coupons and then to components and structures is not a trivial task, since we may have to take into account the effect of different laminate constructions (cross-ply or short fibre), plus the geometric and attenuation effects introduced by dealing with extended components and structures. Crossply and honeycomb materials represent one step towards more realistic materials, whilst model systems (often small pressure vessels) provide some simulation of structural applications.

2.3.4. Component testing

Acoustic emission has been used in the laboratory to evaluate the performance of GRP tubes under conditions of creep and fatigue. This approach has been extended to provide a screening method for GRP tubes designed for use as rocket launchers. A large quantity of effort has also been targeted at testing model pressure vessels, as a halfway stage between simple materials testing and structural monitoring. (This parallels the effort on inspecting pressurised components in the metals field.) It should be noted that such data are not directly relevant to the CARP code on tanks and vessels, as that code covers only low pressure situations. Thus Bunsell, for example, shows that the absence of high amplitude events during fatigue of CFRP pressure vessels indicates structural integrity [52].

Other aerospace-oriented applications include the assessment of missile fins, compression panels and composite wing segments. There are also applications in the automobile industry as a result of the increasing use of composite materials. These include the investigation of tyres, the inspection of GRP springs used in light commercial vans and the testing of sheet moulding compound (SMC) [53].

2.3.5. Structural monitoring

In this section we will consider the application of AE to composite structures. One of the earliest practical applications of AE to composite materials was the Polaris rocket motor case; although the equipment used was relatively crude, the problem and its solution still have relevance today. The problem was that these vessels were designed to a minimum weight criterion, which resulted in their having an effective fatigue life of typically 1 or 2 load applications. Thus a conventional proof test approach could not be used, since vessels that had passed the proof test might be so severely damaged that they would not survive being used. Yet with the less well controlled composite production techniques of the 1960s there were vessels that would fail the proof test and hence would have failed in use. As one of the engineers said at the time, 'virginity was no proof of virtue'. The solution to this impasse was to reduce the pressure test to 60% of the working load; assuming a K^4 damage law, this results in an order of magnitude reduction in damage. Thus all the vessels could be assessed, with minimum damage, and the use of AE allowed the cases to be graded into acceptable and sub-standard units [54].

Another interesting application from the aerospace sector is the F111, although not a fibre-reinforced plastic (FRP) application. This involves detecting the corrosion and debonding which tended to occur at the aluminium/phenolic interface in the leading edge panel in the vertical stabiliser. Water ingress would attack this bond and also give rise to corrosion. The fin was interrogated by thermal means, using a hot air blower. This not only gave rise to a peeling stress at the interface, which tested the bond but also increased the rate of corrosion to a level at which it could be detected by AE. Thus, by using a simple single-channel AE system, this inspection could be carried out rapidly and economically. This AE procedure made significant savings compared with the X-ray technique that had been used previously [55].

In the USA two practical applications of especial interest have arisen; the first of these is 'bucket truck' monitoring. These are inspection vehicles for high voltage transmission lines and consist of a small platform (bucket) which is raised up to the line on a boom. This boom has to be non-conducting and thus is fabricated from GRP. Because of some failures, an inspection method was required and AE was adopted. This test involves extending the boom, and loading it by filling plastic dustbins with water to simulate the weight of the crew. This application is the subject of ASTM standard F914–85 [56].

Perhaps the most important application for AE is for GRP storage tanks and piping [21, 22]. These have been subject to codification by the CARP (Committee on Acoustic emission monitoring of Reinforced Plastics), an SPI committee. The committee's recommended practice for the AE testing of fibreglass-reinforced plastic tanks and vessels was published in 1982, and further codes are in preparation. (The current status of the various standards and codes of practice are summarised in a recent ICE monograph [57].) The code for the tank tests calls for:

(i) pre-test conditioning;
(ii) appropriate sensor layout;
(iii) programmed load with holds and Felicity checks;
(iv) quality judgement based on several aspects of the AE response.

The keys to the success of this code are:

(a) The size of the data base used to set the acceptance criteria (1500 vessels in this case).
(b) The test procedure is designed with AE in mind. Thus we have holds,

Kaiser checks and pre-test conditioning built into the test schedule. Often AE is used as an auxiliary test, monitoring a pressure test that has already been specified. Such a test is unlikely to be optimised for AE testing.

(c) The test reflects the complexity of composite materials and there are several rejection criteria corresponding to the various different degradation mechanisms. These cover:

  (i) *Emission during load hold periods* (< 2 min). If the AE activity does not quickly die down during a hold period at constant load, this is an indication of a lack of structural integrity. (Essentially this shows that some form of creep process is active.)
  (ii) *Felicity ratio* (< 0.95). The Felicity ratio is defined as follows:

$$FR = \frac{\text{Load at which AE activity starts}}{\text{Load previously applied}}$$

The Felicity ratio indicates the presence of damage that had occurred previously. (In this case the AE activity is generated as a result of damage, for example due to frictional sources.)
  (iii) *Total count.* The total count level gives an overall indication of the damage incurred during the load application. Although vessels are rarely rejected on the total count level, it does provide an indication of fabrication quality.
  (iv) *High amplitude events* (< 10). These are a measure of high energy microstructural failure (e.g. fibre failure). This is more often a cause of rejection for pressure vessels than for atmospheric or vacuum vessels.
  (v) *Long duration events.* These events are a measure of macrostructural failure (e.g. delamination).

The key to the success of all these applications is twofold: firstly, the freedom to configure the test specifically to take advantage of the capabilities of AE; and secondly, the development of the accept/reject criteria on a solid basis of practical experience. For example, the CARP code for tanks is based upon the results of testing some 1500 storage vessels. The final factor is the AE signature of the materials involved.

2.3.6. QA and process monitoring

AE has great potential in the field of composite production although as yet there is relatively little published work on this application. One interesting example comes from Dow Chemicals, who are using AE in their promotional literature to 'show' that their resin is especially well matched to glass fibres for composite production. Hamstad [43] has shown that even relatively simple RMS analysis can identify changes in mechanical performance resulting from variations in the process conditions or from post-production environmental attack. Houghton *et al.* [58] have used AE to monitor the curing and cooling cycles of the manufacturing process itself. This has the advantage that no mechanical stress is required to perform the inspection. This approach is extended by the stress-wave-factor (SWF) technique. This method involves:

(a) the injection of an ultrasonic signal into the sample;
(b) the subsequent detection of the signal after its passage through the material;
(c) processing and analysis of the signal, using AE methods.

In effect this is making a measure of the acoustic attenuation of the material. As the state of the material changes, the SWF is also found to vary. Hence the method is being applied to cure monitoring. As the matrix cross-links, the attenuation falls with an associated rise in SWF, which thus provides a measure of the degree of cure and a possible feedback control signal [59]. The SWF can also be applied to post-production and in-service inspection. In these circumstances we need to use a control sample to set the acceptance criteria, and then to use the SWF method to identify variations from this standard. Vary and Lark [60] show how changes in mechanical properties result in changes in SWF. Rodgers [61] concentrates on the use of the SWF method for detecting impact damage. Since such damage is localised, it is necessary to scan the material to find the local anomalies. This scanning is greatly aided by the use of wheeled roller probe sensors, which can be scanned over the surface without the need for any couplant.

2.4. CONCLUSION

This chapter has described the basic technology of acoustic emission and its applications in the composites field. It is evident to those of us working with the technique that the method is becoming more widely used, to meet the more stringent testing and quality assessment requirements which are currently being applied. These are tasks for which the other NDT methods are less well suited. There is still the requirement for continuing research and development effort to help us to correlate AE activity more closely with materials deformation; but the method has clearly demonstrated its practical use to the composites industry in structural monitoring, component QA and more recently in process monitoring. This is a trend that we can expect to continue.

For those readers who wish to find out more about acoustic emission and its application to testing reinforced composites, the proceedings of the first international symposium provide a useful yet relatively compact source, with papers covering materials characterisation, failure mechanisms, evaluation criteria, defect signatures, instrumentation and applications. The book also includes a helpful bibliography listing over 300 references, which provides easy access to more specific areas of interest [62]. The second symposium, held in Montreal in July 1986, also provides a useful supplementary set of proceedings and includes a further bibliography of 154 references [63].

REFERENCES

1. T. F. Drouillard, edited F. J. Laner, *Acoustic emission – a bibliography with abstracts*, IFI/Plenum, New York–Washington–London, 1979.
2. H. N. G. Wadley, C. B. Scruby, D. Stockham-Jones, K. L. Rusbridge, J. H. Worth and J. A. Hudson, *Acoustic emission during deformation and fracture of aluminium alloys: uniaxial tests*, AERE Harwell report AERE-R10362, December 1981.
3. R. M. Belchamber, D. Betteridge, Y. T. Chow, T. Lilley, M. E. A. Cudby and D. G. M. Wood, Looking for patterns in acoustic emission, *1st Int. Symp. Acoustic Emission from Reinforced Composites*, SPI, San Francisco, July 1983, Session 1, 8:55, pp. 1–5.
4. M. Arrington, From Kaiser to Felicity, *Acoustic Emission Trends* (published by *AET*), 1982, **3**(1), 5–6.
5. T. J. Fowler, Acoustic emission testing of glass fibre reinforced composites: state-of-the-art, *Int. Corrosion Forum*: '*Corrosion 82*', NACE, Houston, March 1982, Paper 251.
6. H. Kolsky, *Stress Waves in Solids*, 2nd edition, Dover, New York, 1963, ISBN 0-486-61098-5.
7. W. N. Reynolds and S. J. Wilkinson, The propagation of ultrasonic waves in CFRP Laminates, *Ultrasonics*, May 1974, **12**(3), 109–114.
8. A. M. G. Glennie and J. Summerscales, Acoustic emission source location in orthotropic materials, *British Journal of NDT*, January 1986, **28**(1), 17–22.
9. R. Stiffler and E. G. Henneke II, The application of polyvinylidene fluoride as an acoustic emission transducer for fibrous composite materials, *Materials Evaluation*, July 1983, **41**(8), 956–960.
10. J. C. Wade, P. S. Zerewekhn and R. O. Claus, Detection of acoustic emission in composites by optical fibre interferometry, *Ultrasonics Symposium*, IEEE, Chicago, 1981, pp. 849–852; NASA-CR-164 916, October 1981, pp. 17–23; available from NTIS as N82-11125.

11. D. S. Dean and L. A. Kerridge, *An immersion technique for the detection of stress wave emission in carbon fibre reinforced plastic pressure vessels*, Rocket Propulsion Establishment memorandum RPE-MEMO-672, January 1976. BR 52031.
12. E. V. K. Hill and D. M. Egle, A reciprocity technique for estimating the diffuse-field sensitivity of piezoelectric transducers, J. Acoustical Society of America, February 1980, **67**(2), 666–672.
13. S. L. McBride and T. S. Hutchison, Helium gas jet spectral calibration of acoustic emission transducers and systems, *Canadian J. Physics*, 1 September 1976, **54**(17), 1824–1830.
14. C. C. Feng, *Acoustic emission transducer calibration – spark impulse method*, Endevco Engineering report 74-7-C, 1974.
15. F. R. Breckenridge, C. E. Tschiegg and M. Greenspan, Acoustic emission: some applications of Lamb's problem, *J. Acoustical Society of America*, March 1975, **57**(3), 626–631.
16. A. Nielsen, *Acoustic emission source based on pencil lead breaking*, Svejsecentralen/Danish Welding Institute publication 80.15, København, 1980.
17. M. Arrington, In-situ acoustic emission monitoring of a selected node in an offshore platform, in: M. Onoe, K. Yamaguchi and H. Takahashi (eds), *Progress in Acoustic Emission II*, Proc. 7th International Acoustic Emission Symposium, Japanese Soc. NDI, Zao, October 1984, 381–388.
18. R. E. Green, Jr and J. C. Duke, Jr, Ultrasonic and acoustic emission detection of fatigue damage, *International Advances in NDT*, 1979, **6**, 125–177 (published by Gordon & Breach Science Publishers Inc, USA).
19. A. Vary and K. J. Bowles, Ultrasonic evaluation of the strength of unidirectional graphite polyimide composites, *Proc. 11th Symposium on Nondestructive Evaluation*, San Antonio, Texas, April 1977. NASA-TM-X-73646, 1977, N77-23210.
20. J. Summerscales and D. Short, *An author index to the world literature on acoustic emission in fibre reinforced composite materials*, Plymouth Composite Materials Research Groups bibliography PCMRG-BIB-008, July 1984; available from NTIS as PB85-243 526.
21. C. H. Adams and CARP Committee C on Corrosion Resistant Structures of SPI Reinforced Plastics/Composites Institute, Recommended practice for acoustic emission testing of fibreglass reinforced plastic tanks/vessels, *37th Ann. Tech. Conf.*, SPI, Washington, DC, January 1982, session 27A, 1–3.
22. Discussion by M. A. Droge, Recommended practice for acoustic emission testing of reinforced thermosetting resin pipe (RTRP), *1st Int. Symp. Acoustic Emission from Reinforced Composites*, SPI, San Francisco, July 1983, session 4, 3: 25, pp. 1–17.
23. *Proc. 1st Int. Symp. Acoustic Emission from Reinforced Composites*, SPI, San Francisco, July 1983.
24. *Proc. 2nd Int. Symp. Acoustic Emission from Reinforced Composites*, SPI, Montreal, July 1986.
25. D. O. Harris, A. S. Tetelman and F. A. I. Darwish, *Detection of fibre cracking by acoustic emission*, Dunegan Research Corp technical report DRC-71-1, February 1971.

26. E. G. Henneke II and G. L. Jones, Description of damage in composites by acoustic emission, *Materials Evaluation*, July 1979, **37**(8), 70–75.
27. J. H. Speake and G. J. Curtis, Characterization of fracture processes in CFRP using spectral analysis of the acoustic emission arising from the application of stress, *Proc. 2nd Int. Conf. on Carbon Fibres*, Plastics Institute, London, February 1974, Paper 29, pp. 186–193.
28. J. T. Barnby and T. Parry, Acoustic emission from notched glass fibre reinforced polymer in bending, *J. Physics D: Applied Physics*, 11 September 1976, **9**(13), 1919–1926.
29. L. Golaski, D. Hull and M. Kumosa, Acoustic emission from filament wound pipes under long term loading conditions, *Proc. 4th Int. Conf. Mechanical Behaviour of Materials*, Stockholm, August 1983; Pergamon, 1984, Vol. 1, Sect. 3 (design), pp. 557–563.
30. C.-G. Gustafson and B. R. Selden, Monitoring fatigue damage in CFRP using acoustic emission and radiographic techniques, *ASTM-STP-876*, October 1985, pp. 448–464; *Proc. Symp. Delamination and Debonding of Materials*, ASTM, Pittsburgh, November 1983.
31. R. F. Dickson, C. J. Jones, B. Harris, H. Reiter and T. Adam, Effects of moisture on high performance laminates, *1st Int. Symp. Acoustic Emission from Reinforced Composites*, SPI, San Francisco, July 1983, Session 1, 10:25, pp. 1–14.
32. G. D. Sims, G. D. Dean, B. E. Read and B. C. Western, *Assessment of damage in GRP laminates by stress wave emission and dynamic mechanical measurements*, National Physical Laboratory report NPL-DMA-APP-42, July 1976.
33. B. Harris, F. J. Guild and C. R. Brown, Accumulation of damage in GRP laminates, *J. Physics D: Applied Physics*, 14 August 1979, **12**(8), 1385–1407.
34. M. K. Sridhar, I. Subramaniam, C. Ajay and A. K. Singh, Acoustic emission monitoring of Iosipescu shear test on glass fibre–epoxy composites, *J. Acoustic Emission* Supplement, April-September 1985, **4**(2/3), 174–177; *Proc. 2nd Int. Conf. Acoustic Emission*, Lake Tahoe, October 1985.
35. R. Rothwell and M. Arrington, Acoustic emission and micromechanical debond testing, *Nature Physical Science*, 25 October 1971, **233**(43), 163–164.
36. L. Carlsson and B. Norrbom, Acoustic emission from graphite/epoxy composite laminates, with special reference to delamination, *J. Materials Science*, August 1983, **18**(8), 2503–2509.
37. F. X. de Charentenay, K. Kamimura and A. Lemascon, Fatigue determination in CFRP composites: study with acoustic emission and ultrasonic testing, in: W. W. Stinchcomb (ed.), *Proc. Conf. Mechanics of Nondestructive Testing*, Virginia Polytechnic Institute and State University, September 1980; Plenum, New York, 1981, pp. 391–402.
38. J. E. Flitcroft and R. D. Adams, A study of shear crack propagation in glass and carbon fibre reinforced plastics using acoustic emission monitoring, *J. Physics D: Applied Physics*, 14 June 1982, **15**(6), 991–1005.
39. J. H. Williams and S. S. Lee, Acoustic emission from graphite–epoxy composites containing interlaminar paper inclusions, *NDT International*, February 1979, **12** (1), 5–7.

40. M. Wevers, I. Verpoest, E. Aernoudt and P. de Meester, Analysis of fatigue damage in carbon fibre reinforced epoxy composites by means of acoustic emission: setting up a damage accumulation theory, *J. Acoustic Emission* Supplement, April-September 1985, **4**(2/3), 186–190; *Proc. 2nd Int. Conf. Acoustic Emission*, Lake Tahoe, October 1985.
41. A. R. Bunsell and B. Harris, Hybrid carbon and glass fibre composites, *Composites*, July 1974, **5**(4), 157–164.
42. M. P. Ansell, Acoustic emission from softwoods in tension, *Wood Science and Technology*, 1982, **16**(1), 35–58.
43. M. A. Hamstad, Acoustic emission quality control of Kevlar 49 filament wound composites, *12th National Technical Conference*, SAMPE, Seattle, October 1980, pp. 380–393.
44. D. Betteridge, P. A. Connors, T. Lilley, N. R. Shoko, M. E. A. Cudby and D. G. M. Wood, Analysis of acoustic emissions from polymers, *Polymer*, September 1983, **24**(9), 1206–1212.
45. A. A. Pollock, Acoustic emission 2: acoustic emission amplitudes, *Non-Destructive Testing*, October 1973, **6**(5), 264–269.
46. K. Wolitz, W. Brockmann and T. Fischer, Evaluation of glass fibre reinforced plastics by means of AE measurements, *4th Acoustic Emission Symposium*, Japanese Soc. NDI, Tokyo, September 1978, Paper 5-4, pp. 5.52–5.58.
47. M. Eikelboom, *Acoustic emission of carbon fibre reinforced plastics*, Fokker report FOK-R-2464, 29 August 1979; available from NTIS as N81-12250.
48. F. J. Guild, M. G. Phillips and B. Harris, Acoustic emission studies of damage in GRP, *NDT International*, October 1980, **13**(5), 209–218.
49. J. H. Williams, Jr, and D. M. Egan, Acoustic emission spectral analysis of fibre composite failure mechanisms, *Materials Evaluation*, January 1979, **37**(1), 43–47.
50. L. J. Graham, Acoustic emission signal analysis for failure mode identification, *Spring Conf.*, ASNT, Philadelphia, March 1980, 74–79.
51. E. G. Henneke II, *Signature analysis of acoustic emission from composites*, NASA report NASA-CR-145 373, 19 May 1978; available from NTIS as N78-23148.
52. A. R. Bunsell and M. Fuwa, An acoustic emission proof test for a carbon fibre reinforced epoxy pressure vessel, *8th World Conf. on NDT*, Cannes, September 1976, Paper 4B/6.
53. D. L. Sherer and M. Ashley, Acoustic emission testing of HMC composites, *36th Ann. Reinforced Plastics/Composites Institute Tech. Conf.*, SPI, Washington, DC, February 1981, Paper 23B.
54. A. T. Green, C. S. Lockman and R. K. Steele, Acoustic verification of structural integrity of Polaris chambers, *Modern Plastics*, July 1964, **41** (11), 137–139/178/180.
55. J. R. Rodgers and S. Moore, *The use of acoustic emission for detection of active corrosion and degraded adhesive bonding in aircraft structures*, McClellan Air Force Base (CA) report, 1976.
56. *Standard test method for acoustic emission for insulated aerial personnel devices*, ASTM standard designation F914-85.

57. *Guidance notes on the use of acoustic emission tests in process plants*, 3rd report of the Internal Study Group on Hydrocarbon Oxidation of the Institute of Chemical Engineers, London, 1985.
58. W. W. Houghton, R. J. Shuford and J. F. Sprouse, Acoustic emission as an aid to investigating composite manufacturing processes, *11th Nat. Tech. Conf.*, SAMPE, Boston, November 1979, 131–150.
59. R. J. Hinrichs and J. M. Thuen (Applied Polymer Technology Inc), Control system for processing composite materials, *United States Patent 4 455 268*, 19 June 1984.
60. A. Vary and R. F. Lark, Correlation of fibre composite tensile strength with ultrasonic stress wave factor, *J. Testing & Evaluation*, July 1979, **7**(4), 185–191.
61. J. M. Rodgers, Advances in acousto-ultrasonic inspection of composites and adhesively bonded structures, In: M. Onoe, K. Yamaguchi and T. Kishi (eds), *Progress in Acoustic Emission I, Proc. 6th Int. Acoustic Emission Symposium*, Japanese Soc. NDI, Susono, October/November 1982, 323–331.
62. T. F. Drouillard and M. A. Hamstad, A comprehensive guide to the literature on acoustic emission from composites, *Proc. 1st Int. Symp. Acoustic Emission from Reinforced Composites*, SPI, San Francisco, July 1983, Session 6, 10:25, pp. 1–60.
63. T. F. Drouillard, A comprehensive guide to the literature on acoustic emission from composites, *Proc. 2nd Int. Symp. Acoustic Emission from Reinforced Composites*, SPI, Montreal, July 1986, Session 2, 4:10, pp. 60–71.

Chapter 3

Thermal NDT Methods

K. E. PUTTICK
Department of Physics, University of Surrey, Guildford, UK

3.1. INTRODUCTION

Composite materials present a number of novel problems compared to traditional engineering metals. Besides surface cracks and holes, there may occur subsurface delaminations or other laminar flaws; in addition, the increasing use of short fibre composites to fabricate load-bearing components of complex shape implies the presence of potentially harmful anomalies such as variations in fibre alignment (leading to weakness in the direction perpendicular to the preferred direction) or changes in fibre density. Since all such inhomogeneities are associated with changes in thermal properties, the use of various ways, now widely available, of visualizing heat flow provides rapid and extremely versatile imaging of defect sites.

3.1.1. Active heating

Much of the early work on thermal imaging of composite and other materials was devoted to the so-called active method or stress-generated thermal field (SGTF) (e.g. McLaughlin *et al.* [1]). This makes use of the fact that under cyclic loading heat is generated in regions of damage by various mechanisms: for example, the rubbing together during the compression cycle of the two faces of a fatigue crack, or by hysteresis within the matrix material. These hot spots are particularly evident in specimens, such as GRP, of low thermal conductivity [2–5]. The commonest type of experiment involves fatigue stressing in conventional frequency ranges, but Pye and Adams [5] have emphasized the use of resonant vibration as a means of maximizing the effect of a given applied force. Since the temperature rise is proportional to the applied frequency, there may also be a considerable advantage in employing ultrasonic frequencies [4].

The advantage of active heat generation NDT is that the grossest defects in mechanical properties of a structure may be immediately displayed without recourse to lengthy scanning or point-to-point examination. The disadvantages are first, the need to attach the specimen to some kind of cyclic loading device, which limits the shape and size to some extent, and second, the danger that the loading itself may actually extend the damaged area under investigation and so by definition cease to act as a non-destructive test. However, it has been found that in favourable cases cracks of < 14 mm can be detected in GRP; in CFRP the minimum crack size is considerably higher, and defects may often be unresolved due to the greater thermal conductivity of the material.

Recently a remarkable new tool of the active heating type has become available, based on a highly sensitive infra-red camera capable of detecting changes of 10^{-3} deg K in regions less than 1 mm, for the quantitative evaluation of stress fields. This system, known as SPATE (Stress Pattern Analysis by Thermal Emission), visualizes temperature changes due to the thermoelastic effect [6]. The signals received by the detector are correlated in frequency, magnitude and phase with a reference signal derived from a load transducer; the correlation rejects extraneous radiation, and gives an output which is a function of stress and whose sign identifies tensile or compressive components. The display on the CRT monitor is analogous to the fringe pattern observed in conventional techniques.

At present the method is truly quantitative only for isotropic materials such as metals, and an extension of the mathematical analysis is required for materials such as composites whose thermal as well as mechanical properties are highly anisotropic.

However, it can still be used qualitatively as a vivid way of locating defects, for example the tip of a crack, and for monitoring the spread of damage. SPATE, though at present a somewhat expensive procedure justifiable only in crucial types of inspection, clearly holds much promise as a means of identifying potentially harmful inhomogeneities in composites.

3.1.2. Passive heating

The alternative mode of thermal inspection involves the application of a local or general source of heat followed by observation of the subsequent

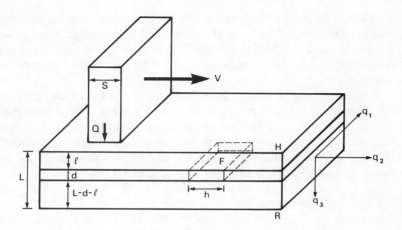

Fig. 1. General scheme for thermal non-destructive testing.

temperature distribution (the 'passive' or externally applied heat field method). This class of experiment has become increasingly popular with the advent of commercial systems affording sophisticated visual displays which can be recorded for subsequent analysis.

In general, the method relies on the fact that deleterious anomalies in materials perturb the heat flow and therefore the temperature field characteristic of a homogeneous solid. Observations are naturally most sensitive which relate to surface defects, but by careful attention to experimental conditions it has been shown that interior flaws may also be rendered visible, particularly at laminar interfaces or adhesive bonds. This chapter concentrates on the use of passive thermography (otherwise known as the use of externally applied thermal fields or EATF) with infrared scanners or liquid crystals.

3.2. EXPERIMENTAL METHODS IN PASSIVE HEATING

3.2.1. Heat sources

The usual methods of generating the required heat flux involve uniform heating of the specimen surface on one side and subsequent detection of the temperature field either on the same side or the obverse during cooling. A typical source is a bank of lamp bulbs controlled by some timing device; recently photographic flash tubes have been used successfully to produce a well-defined pulse of thermal energy [7]. Hot liquids and air dryers have also been used in particular applications, and where elevated temperatures cannot be used it is sometimes possible to cool the test piece rapidly by quenching or a freezer spray and record the resulting temperature rise to ambient.

These procedures tend to limit the spatial resolution compared with that attainable by infra-red detecting systems. This may be important in the investigation of short fibre composites fabricated by injection moulding, where material properties may vary significantly over distances of the order of 1 mm, and attention has therefore been directed to the use of point sources of heat, either static [8] or scanned [9].

3.2.2. Temperature imaging

Recording the transient temperature contours of the specimen may be accomplished either by electronic infra-red detectors or by liquid crystals. The former represents the most usual technique, on account of the salient advantages of completely remote sensing and versatility of signal processing furnished by modern digital procedures. However, as emphasized by Wilson and Charles [10], liquid crystals represent a low cost alternative with high accuracy ($\sim 10^{-2}$ deg K and 1 μm spatial resolution), fast response time and continuous tracking of thermal fields. The main disadvantage of liquid crystals is the necessity of applying them to the specimen surface in a thin uniform layer; the way in which this is carried out may affect the quality of the thermography results. Also, since liquid crystals produce contrast by light scattering, the best results are obtained on black surfaces, so that they are easier to use on CFRP rather than GRP composites.

A somewhat different method, known as photothermal imaging, is based on the use of a periodically modulated scanned point heat source to produce a heavily damped temperature wave. The periodic surface temperature detected by an infra-red scanner will be influenced by the scattering or reflection of the thermal wave when it encounters regions of different thermal characteristics [9]. The technique has been successfully used to monitor bonding defects between substrates and plasma-sprayed coatings. At a coating–substrate interface the coefficient of reflection of the thermal waves changes at a defect, leading to variations in both the detected signal and its phase; recording the latter has the advantage that the phase angle is independent of any optical structure on the coating surface. Such photothermal testing may have applications to the study of interfacial defects in laminated composites.

3.3. THEORETICAL ASPECTS

It has not so far proved possible to make reliable calculations of heat conduction in the vicinity of most real defects, particularly in laminar or short fibre composites, and there remains an urgent need for systematic empirical studies of standard reference specimens containing known defects in these materials. Nevertheless, theoretical analysis, especially of course with numerical calculation of idealized inhomogeneities, remains important in understanding the optimum conditions of temperature and time required for contrast of sub-surface flaws. In this section, therefore, we propose to illustrate the main features of such simulation and their comparison with experimental results.

3.3.1. Numerical modelling

The work of Vavilov and co-workers [11] exemplifies the principles of thermographic computer modelling. The situation analysed is shown in Fig. 1: a strip heat source moving with velocity V scans the surface of a block of material of thickness L and extending indefinitely in the other two dimensions. A square defect of side h is at a depth l below the heated surface. Temperatures are estimated on the heated surface (H) and the reverse side (R), and plotted in the form of the temperature contrast parameter $A = \Delta T/T$, where ΔT is the local temperature rise due to the presence of the defect. The basic equations for numerical solution are:

(i) The transient three-dimensional equation of heat transfer

$$a V^2 t_j = \frac{\partial t_j}{\partial \tau}; \quad j = 1 - n \tag{1}$$

Here j is the layer number, n the number of layers, t is temperature, τ time and a the absorptivity.

(ii) The combined boundary condition at the specimen surface, including heat transfer by radiation and convection

$$\frac{\lambda \partial t_j}{\partial q_i} = -Q + \varepsilon\sigma \left| \left(\frac{t_j}{100}\right)^4 \left(-\frac{t_a}{100}\right) \right| + K(t_j - t_a). \tag{2}$$

where λ is the conductivity, ε the emissivity and σ the Stefan–Boltzmann constant.

(iii) The equation of the heat flux at the interlayer boundary

$$\lambda_j \frac{\partial t_j}{\partial q_i} = \lambda_k \frac{\partial t_k}{\partial q_i} \quad (3)$$

(iv) The equation of continuity

$$t_j = t_k \quad (4)$$

(v) $Q = \begin{cases} Q(q_i, \tau) & \text{inside the heat beam} \\ O & \text{outside the heat beam} \end{cases} \quad (5)$

(vi) The initial condition

$$t_j(\tau = 0) = t_a, \quad (6)$$

where t_a is the ambient temperature.

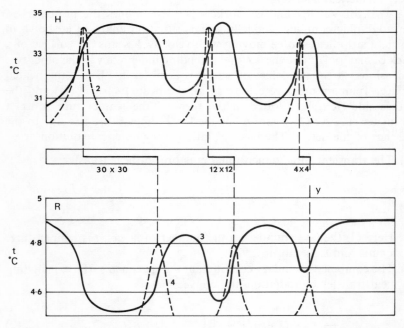

FIG. 2. Temperature profiles (curves 1, 3) and derivatives $\partial t/\partial y$ (curves 2, 4) on H, R surfaces for plastic. $L = 6.1$ mm; $l = 2$ mm; $V = 5.3$ mm/s; $Q = 10$ kW/m²; $S = 40$ mm (line heater); $1 - \tau_0 = 7.2$ s; $3 - \tau_0 = 12.8$ s. Note that the derivative maximum corresponds to a rising temperature signal, the minimum occurs for a decreasing temperature signal.

Detailed solutions have been achieved for the above equations with two simplifications: neglect of the radiation term in (2), and two dimensions only. The calculation grid comprised 16000 points, and results were obtained for a variety of real materials, including metals, plastics and various types of bonded composites. Typical parameter value ranges are: $L = 1-10$ mm, $h = 4-30$ mm, $V = 1-10$ mm s^{-1}, $Q = 7-50$ kW m^{-2}. Defect thickness $d = 0.1-0.3$ mm. Typical results are shown in Fig. 2 in the form of temperature profiles for three sizes of defect 2 mm below the surface of a 6.1 mm thick sheet of plastic (the nature of which is not specified). Also indicated are the first derivatives of the temperature curves, showing that in each case the maxima of these derivatives coincide with the projection of a defect edge; it has been found that this criterion is independent of other parameters and may well be of general use in interpreting thermographic results.

The sensitivity of the technique for a range of materials, defined by the minimum cross-section of detectable defects, is shown in Table 1. The marked dependence on specimen thickness is noteworthy. A further conclusion from the study is that the minimum sizes of delaminations and disbonds should be 2–3 times greater than the thickness of the layer over them. A somewhat different simulated experiment has recently been analysed by Sayers [12]. Calculations were made of the heat flow through a material of rectangular cross-section, assumed to be of infinite length in one direction, containing a linear defect of constant cross-section running down its length. Again, the computation was carried out for the two-dimensional case only, and time dependence was investigated at equal time intervals Δt. The lattice of nodes employed as a basis for the analysis is shown in Fig. 3: each node represents a volume of area with a square cross-section in the x, y plane and unit depth, say, in the z direction. In Sayer's notation temperature is represented by T, so that the temperature in the volume around the node l, m at time interval p is T_{lm}^p. Energy balance in the neighbourhood then gives

$$-k_x \Delta y \frac{(T_{l,m}^p - T_{l-1,m}^p)}{\Delta x} - k_x \Delta y \frac{(T_{l,m}^p - T_{l+1,m}^p)}{\Delta x}$$

$$-k_y \Delta x \frac{(T_{l,m}^p - T_{l,m-1}^p)}{\Delta y} - k_y \Delta y \frac{(T_{l,m}^p - T_{l,m+1}^p)}{\Delta y}$$

$$= \rho_c \Delta x \Delta y \frac{(T_{l,m}^{p+1} - T_{l,mm}^p)}{\Delta t} \text{ for unit depth} \quad (7)$$

TABLE 1

No.	Material and its thickness	Minimum cross-sectional area of detectable defect (mm^2)
	Composites	
1	Plastic–titanium	100
2	Plastic (3 mm) – metal (5 mm) – plastic (3 mm)	100
3	Glass (0.2–1 mm) – Al (1.5 mm)	1
4	Al (2 mm) – asbestos (4 mm)	100
	Steel (2 mm) – rubber (5 mm)	100
5	Al (2 mm) – honeycomb core (20 mm) – Al (2 mm)	140
6	Ferrite (1–4 mm) – metal	0.5–7
7	Metal (6 mm) – plastic (3 mm) – metal (6 mm)	400
8	Fibreglass (3.2 mm) – Al honeycomb core (9.1 mm)	100
	Non–metals	
1	Textolite, Teflon, ceramics (2–10 mm)	1–25
2	Rubber	50–400
3	Wood, plastic (5 mm)	100
4	Fibreglass (0.3 mm)	0.75
5	Fibreglass (7 mm)	225
6	Fibreglass (12 mm)	100
7	Plastic (7 mm) – plastic (45 mm)	100
8	NbC – graphite	10
	Metals	
1	Al–Al, brass–brass, steel–steel	0.1–7
2	Cu (0.3 mm)–Cu	0.5
3	Al, Zircaloy (0.7 mm) – fuel element core	100
4	Titanium (0.5 mm) – Al (2–4 mm)	25
5	Babbit (0.8 mm) – steel (12 mm)	140

If, further, we choose $\Delta x/\Delta y$ such that $k_x(\Delta y/\Delta x) = k_y(\Delta x/\Delta y) = k$, say, giving

$$T_{l,m}^{p+1} = \frac{\alpha \Delta t}{\Delta x \, \Delta y}(T_{l-1,m}^p + T_{l+1,m}^p + T_{l,m-1}^p + T_{l,m+1}^p)$$

$$+ \left(1 - \frac{4\alpha \Delta t}{\Delta x \, \Delta y}\right)T_{l,m}^p \quad (8)$$

where $\alpha = k/\rho_c$.

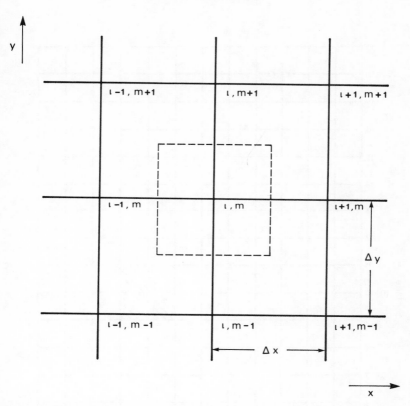

FIG. 3. Division of space into a lattice of nodes, each node representing a volume of area in the xy plane indicated by the dotted lines and of unit depth, say, in the z direction.

Similar equations may be derived for inequivalent nodes at the surface and corners of the specimen as well as in the vicinity of a defect.

The computer code used to solve these equations may be applied to arbitrary specimen shapes and three-dimensional defect geometries. Calculations were actually performed for a 2×2 void at various depths in the 13×19 lattice shown in Fig. 4. If eqn (8) is written in the form

$$T_i^{p+1} = \sum_{j}' \eta_{ij} T_j^p \qquad (9)$$

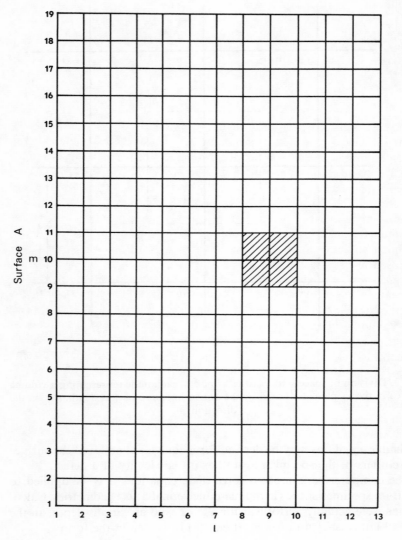

FIG. 4. Typical position considered for a 2 × 2 void in a lattice of dimensions 13 × 19.

where the prime indicates that the sum is restricted to i and its nearest neighbours,

$$\eta_{ij} = \frac{\alpha \Delta t}{\Delta x \Delta y}; i = (l,m), j = (l-1,m)(l+1,m)(l,m-1)(l,m+1)$$

$$= \left(1 - \frac{4\alpha \Delta t}{\Delta x \Delta y}\right) \quad i = j = (l,m)$$

In the case under consideration there are six types of nodes, and only 48 different η_{ij} need be stored. For any given material the real time interval corresponding to a time step may be evaluated: for perspex, for example, with a thermal diffusivity of 1.08×10^{-3} cm^2 s^{-1}, the value of $\alpha \Delta t / \Delta x \Delta y$ was chosen to be 0.1728, giving $\Delta t = 400$ ms for $\Delta x = \Delta y = 0.5$ mm. Heat flow through the specimen cross-section was computed for (i) the 2×2 defects lying at various depths and positions in the specimen, and (ii) an instantaneous temperature rise uniform over an entire surface of the specimen at the beginning of the first time interval. The specimen was supposed to be at a uniform temperature T_0.

Fig. 5 shows a typical set of results of the temperature distribution over the heated surface for defects lying on the mid-line of the slab. It is evident that the contrast falls off rapidly with the depth of the void; it appears also that Vavilov's criterion for determining defect size holds in this situation. This is, in fact, the most favourable side for observation, in the sense that the contrast is greatest: the comparable distribution on the opposite side shows much less contrast as well as, of course, a smaller general temperature rise and a much longer time before the attainment of maximum contrast. Table 2 summarizes the main features of the simulation, indicating that a combination of one-sided and two-sided inspection ('reflection' and 'transmission') allows an observer to distinguish between large defects remote from the heated surface and small defects close to it.

A numerical model specifically constructed for laminated fibre-reinforced composite material was developed by Charles and Wilson [13]. Within each lamina, the thermal conductivities in the longitudinal (fibre) and transverse directions were taken to be respectively

$$\left. \begin{array}{l} k_1 = k_f V_f + k_m (1 - V_f) \\[6pt] k_2 = k_m \left[\dfrac{k_f(1+V_f) + k_m(1-V_f)}{k_f(1-V_f) + k_m(1+V_f)} \right] \end{array} \right\} \quad (10)$$

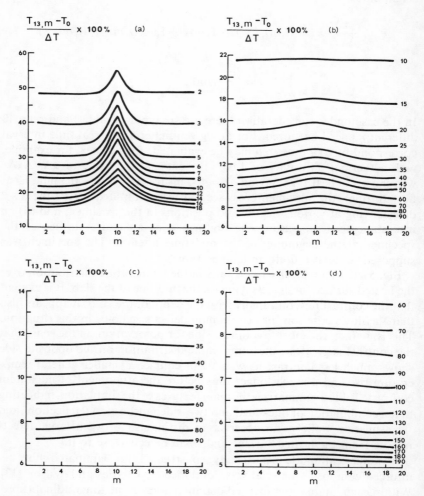

FIG. 5. Temperature distribution T_{13}, m, $m = 1, 19$, over side B at the beginning of the first time interval for various positions of the defect. (a) $l = 11$, $m = 10$; (b) $l = q$, $m = 10$; (c) $l = 7$, $m = 10$; (d) $l = 5$, $b = 10$. The number of time steps having elapsed is indicated on the curves.

TABLE 2

Defect depth below side B	One-sided inspection		Two-sided inspection	
	$\dfrac{(T_{13,1} - T_{13,10})_{\max}}{\Delta T} \times 100\%$	Number of time steps after which maximum contrast occurs	$\dfrac{(T_{13,1} - T_{13,10})_{\max}}{\Delta T} \times 100\%$	Number of time steps after which maximum contrast occurs
$2\Delta x$	11.41 %	6	0.13 %	80
$4\Delta x$	1.08 %	38	0.15 %	104
$6\Delta x$	0.26 %	82	0.12 %	109
$8\Delta x$	0.08 %	122	0.08 %	101
$10\Delta x$	0.02 %	159	0.04 %	88

where $k_f \equiv$ fibre conductivity, $k_m \equiv$ matrix conductivity and $V_f \equiv$ volume fraction of fibre.

The components of conductivity in any given co-ordinate system (x, y, z) are

$$\left.\begin{array}{l} k_x = |k_1 \cos \theta| + |k_2 \sin \theta| \\ k_y = |k_1 \sin \theta| + |k_2 \cos \theta| \\ k_z = k_2 \end{array}\right\} \quad (11)$$

where $\theta =$ fibre orientation angle with respect to the x axis.

At a flaw the conductivity in the z direction was assumed to be given by

$$(k_z)_{\text{defect}} = \text{DCF}\, k_z \quad (12)$$

where $\text{DCF} \equiv$ defect conduction factor with values between 0 and 1. The defects are located between laminae, so that the x and y conductivities are unaffected. Using the parameters so defined, solutions were obtained for the differential equation for anisotropic heat flow

$$\rho C_p \frac{\partial T}{\partial t} = k_x \frac{\partial^2 T}{\partial x^2} + k_y \frac{\partial^2 T}{\partial y^2} + k_z \frac{\partial^2 T}{\partial z^2} \quad (13)$$

where $\rho \equiv$ net material density, $C_p \equiv$ net material specific heat and $T \equiv$ temperature. Using a finite difference procedure, numerical results were obtained for two particular systems, one graphite–epoxy and one fibreglass–epoxy composite. These results were presented in terms of the 'edge temperature gradient', the gradient across the edge of the defect projection on the viewed surface. Three salient features are noteworthy:

(i) The maximum contrast, defined as the peak value of the edge gradient as a function of time, depends markedly on the way in which the applied heat is generated. Radiant heating affords a more gradual decay of the edge gradient with time (thus allowing more time for measurement) than contact heating.

(ii) For comparable specimens (same number of plies, same size of defect), the same heat input yields much higher edge gradients for longer periods with fibreglass than with carbon fibre composites.

(iii) For a given defect conduction factor the peak edge temperature gradient as a function of defect depth has a maximum at a certain depth: for example, for a DCF of 0.5 in CFR–epoxy the optimum depth occurs at three plies from the surface.

An explicit solution was obtained by Williams *et al.* [14] for the one-dimensional heat flow in a flat plate containing a planar flaw and heated on one side by a spatially uniform step heat flux. Each side of the plate was characterized by a uniform heat transfer coefficient h, the matrix by a conductivity k_1, and the flaw by a conductivity k_2. The temperature rises in the unflawed and flawed part of the plate, respectively T_{IS}, T_{IIS} were computed, and the difference between them, ΔT, presented in the form of a non-dimensional parameter

$$\Delta T^* = \frac{k_1 \Delta T}{q'' l} \tag{14}$$

where q'' is the instantaneous applied heat flux and l the plate thickness. As in the previously described computer simulations, ΔT^* is found always to have a maximum or minimum during the course of the test representing the optimum time of observation, and its absolute magnitude is sensitive to the heat transfer coefficient h, the magnitude of the heat flux and the ratio of thermal conductivities of matrix and flaw. Interestingly, ΔT^* is relatively insensitive to the difference in thermal diffusivity between matrix and flaw; it is also noteworthy that this temperature contrast increases as the thickness of the plate decreases.

3.3.2. Comparison of model predictions with experimental results

Detailed results obtained by two finite-difference computer codes have recently been compared with experimental results [15]. These were, first, the program known as 2DT originally developed for nuclear reactor work, and, second, the Sayers program described in the previous section. These were first tested on a problem with an analytical solution (with which the results were compared), that of a simple semi-infinite slab heated by an instantaneous pulse. The 2DT code was then applied to more complex specimens consisting of perspex blocks containing idealized defects at different depths below the heated surface, and its predictions were compared with actual observation of the temperature contrast, due to such defects, using an infra-red TV camera. The 2DT program is versatile and capable of application to many problems directly relevant to the testing of composites, including problems involving different materials, and the anisotropic thermal conductivities and specific heats associated with fibre-reinforced solids; in addition, radiative and convective losses from specimen boundaries can, if necessary, be incorporated in the finite-difference equations.

The particular material chosen for comparison of these codes with an analytic solution was perspex. For the calculation it was assumed that at time $t = 0$ the temperature of the semi-infinite slab specimen of thickness l was zero except in a narrow region $0 \leq x \leq \varepsilon (\varepsilon \ll l)$ where the temperature was T_0. For $t > 0$ the heat from the narrow region diffuses into the remainder of the slab. The analytic solution for the problem is [16]:

$$T(x, t) = T_{\mathrm{f}}\left[1 + \frac{2l}{\varepsilon\pi} \sum_{n=1}^{\infty} \frac{1}{n} \sin\frac{n\pi x}{l} \cos\frac{n\pi x}{l} \exp\left(\frac{-n^2 \alpha^2 \pi^2 t}{l^2}\right)\right] \quad (15)$$

where $\alpha^2 = k/\rho c$.

For perspex, $\alpha = 1.08 \times 10^{-3}$ cm^2 s^{-1}. l_1 was taken to be 6 mm, $\varepsilon = 10^{-3}\, l$. The initial time steps were 0.04 s, 0.23 s for the 2DT and Sayers programs respectively. Figures 6 and 7 show respectively the temperatures of the heated and non-heated surfaces of this semi-infinite perspex slab as a function of time as calculated from the analytic solution and the two finite-difference programs. In Fig. 6 it is seen that for small times there is significant divergence of the Sayers results from the analytical solution due to the larger time step, but that for times > 3 s there is good agreement between the three solutions. On the other side of the slab there is almost perfect agreement between the Sayers predictions and the analytical solution because there are many time steps until the initial temperature rise. The 2DT temperatures are significantly lower, probably because of small heat losses from the model boundaries.

The 2DT program was compared with observation for the case of two perspex blocks containing 1 mm diameter holes through the width direction (Fig. 8). In the experiment, pulses of heat were applied to the block surfaces and the infra-red television camera was used to monitor the temperature of the same surface as a function of time. The 2DT code modelled this situation with the following simplified assumptions:

(a) The perspex block extends indefinitely in the z direction, so that the problem is essentially two-dimensional.
(b) Heat losses from the boundaries of the block can be neglected.
(c) The holes have a square rather than (as in the actual case) circular cross-section.
(d) Thermal diffusivity is isotropic and temperature independent.
(e) The heat pulse is instantaneous at $t = 0$.

Results were obtained in the form of the difference ΔT between the surface temperatures directly over each hole and the general uniform

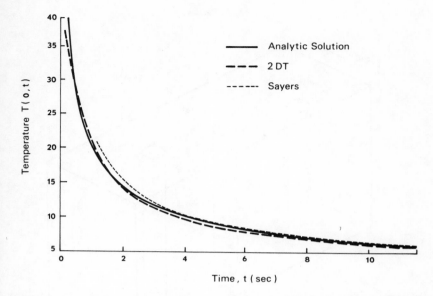

FIG. 6. Temperature of the heated face of a semi-infinite perspex slab as a function of time, as calculated from the analytical solution, the 2DT and the Sayers programs.

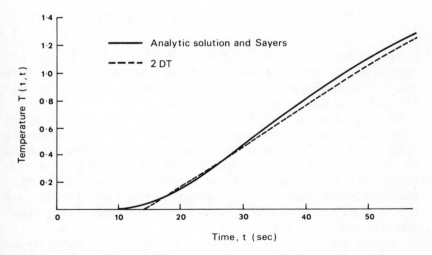

FIG. 7. Temperature of the non-heated face of a semi-infinite perspex slab as a function of time, as calculated from the analytic solution and the two finite-difference computer programs.

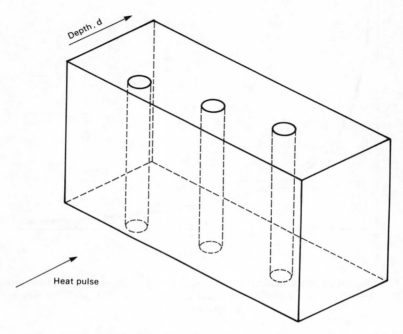

FIG. 8. Perspex test block used for transient thermography experiments.

surface temperature well removed from the holes. Figure 9 shows predictions of this contrast for three holes at different depths in a 6 mm thick block. As is seen in the figure and in Table 3, the maximum contrast falls off rapidly and its time increases with depth of hole.

These curves were compared with the temperatures recorded by the monitor on videotape and subsequently analysed by image processing methods. The absolute values of ΔT obtained from the models cannot, however, be directly compared with the experimental values because of uncertainties in the amount of heat supplied to the specimen and in emissivity effects. The observed values are therefore scaled to form the best fit to the model predictions so as to verify the trends of temperature with time and hole depth.

The experimental results are significantly low, and a separate investigation suggested that an important source of error lay in an overestimate of the cross-sectional area of the hole. However, the model gave better agreement with the experimental values for a specimen containing five

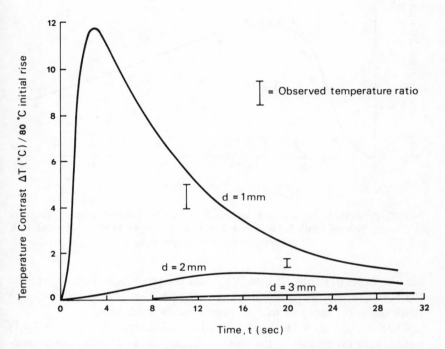

Fig. 9. Predictions (solid lines) from the 2DT code for the temperature contrasts of 1 mm square holes with mean depths, $d = 1, 2$ and 3 mm in a 6 mm thick Perspex block. Two experimental points for the $d = 1$ mm hole are shown for comparison with theory.

TABLE 3

Results from the 2DT Code for Maximum Temperature Contrast for 1 mm Square Holes in a 6 mm Perspex Block

Mean hole depth (mm)	Time at which maximum contrast occurs (s)	Maximum contrast (deg)
1	2.8	11.9
2	15.5	1.11
3	27.5	0.26

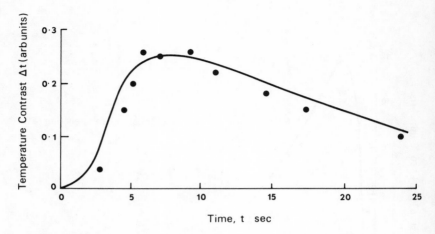

Fig. 10. Predictions (solid line) from the 2DT code for the temperature contrasts of a 1 mm hole of depth 1.5 mm in a 3 mm thick perspex block, compared with observational results (points).

circular holes of depth 1.5 mm in a 3 mm perspex block: Fig. 10 presents the 2DT temperature–time curve together with a set of observations, each point being the mean of five readings (one for each hole).

Charles and Wilson [13] used a liquid crystal detector (see section 3.4.1) to verify the predictions of the numerical calculations of heat flow in fibre-reinforced laminate composites. Good agreement between the calculated and observed temperature rises on both the front and back surfaces was found.

3.4. EXPERIMENTAL WORK

3.4.1. Liquid crystal recording

Although scanning infra-red imaging undoubtedly represents the most convenient and genuinely remote sensing method available for thermographic inspection, excellent results have been obtained at low cost by means of liquid crystal (LC) imaging (sometimes known as chemical imaging, though strictly speaking no chemical reaction takes place). LC thermography was applied early to evaluation of composite materials [17], and more recently Charles and co-workers [10, 18] have made extensive

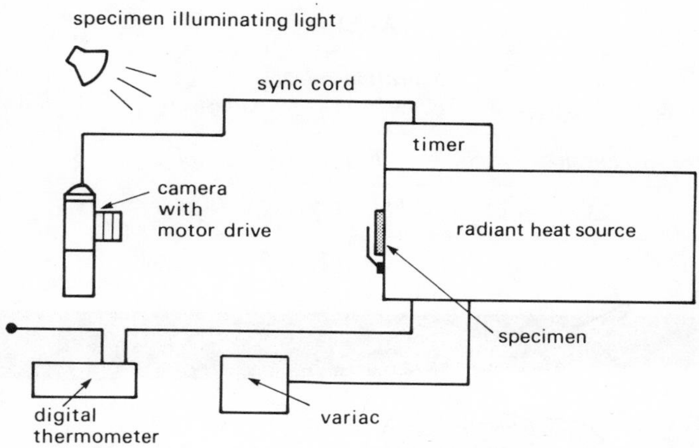

FIG. 11. Experimental arrangement used by Wilson and Charles [10] (courtesy Society for Experimental Mechanics).

use of the technique for detection of adhesive-bond-line and interlaminar flaws in a variety of composites, and compared the images with those obtained by ultrasonic C-scan measurements and by direct microscopic examination of sectioned specimens.

The experimental arrangement used by Wilson and Charles [10] is delineated in Fig. 11. The radiant source was a bank of eight 100 W bulbs in a shielded enclosure, heating the back surface of a specimen, and a low speed (3 frames/s) cine camera used to record the thermal images in colour film. A digital clock indicating time elapsed from the start of radiant heating was included in each photograph.

The liquid crystal compound used in the investigation had a 'first-indication' temperature at which red light was first scattered of 24 °C, and the range of the compound was 3 °C, i.e. the crystals became clear at 27 °C. It was applied to the surfaces of the composites by spraying a thin uniform layer on areas previously cleaned by petroleum ether and spray-painted flat black to enhance image quality. Test specimens were initially brought to equilibrium 2 °C below the first-indication temperature.

The compound used displays four indication temperatures: red, yellow, green and blue. These were calibrated against two independent thermometric measurements, which revealed significant non-linearity in the temperature ranges associated with each colour (Table 4).

The calibration was found to be valid only for a limited period of several

TABLE 4

Red		Indication colour Yellow		Green		Blue	
Start	Finish	S	F	S	F	S	F
°C 22.8	23.5	23.5	24.3	24.3	25.5	25.5	28.4

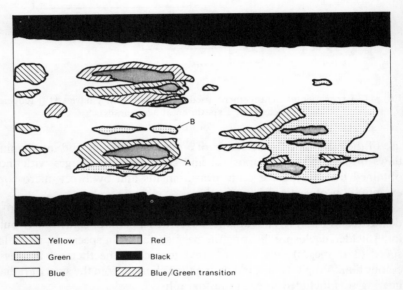

FIG. 12. Schematic mapping of LC thermogram at $t = 6.17$ s for carbon fibre–epoxy bonded joint.

hours; this was, however, sufficient for complete tests to be repeated with consistent results. Thermographic observations made in this way were compared with pulse-echo ultrasonic C-scans with gated peak-height analysis and with low power (50×) optical microscopy on sections of specimens in the flawed regions.

A typical thermograph is shown in Fig. 12 with the original colours represented by shading. This figure, of a carbon fibre–epoxy bonded joint, includes two areas marked A and B which have maximum temperature differences with unflawed surroundings of 2° and 1° respectively; subsequent microstructural analysis revealed that flaw A was approximately 1.14 mm thick as against a thickness of 0.61 mm for flaw B. Such thermograms in general agreed well with the results of C-scan and microscopic study. Lee and Williams [19] used the one-dimensional heat flow equation described in Section 3.3.1 to interpret their observations of temperature distribution in transient heating of flawed fibreglass composites by a liquid crystal method. Their results agreed well with the theoretical predictions of the non-dimensional surface temperature defined in Section 3.3.1, and the effectiveness of the test was analysed in terms of the number of colour differences observed in the region of the projection of the flaws on the surface. Four ranges of colour difference numbers were used to characterize the results: 0–1, 1–2, 2–6 and > 6, labelled respectively 'unacceptable', 'limiting', 'moderately acceptable' and 'acceptable'. A sample of the observations, relating depth of flaw to number of colour differences, is given in Fig. 13.

Reynolds and Wells [7] used a scanning infra-red camera in conjunction with a television display to detect delaminations, disbonds and other flaws in a variety of composites and bonded structures. A heat source in the form of a vitreosil–xenon photographic flash tube was used to deliver a pulse of radiant heat to the whole of one surface of the specimen, and the subsequent rise and decay of temperature on either the heated or the opposite face was recorded on magnetic tape and displayed with facilities for reverse running, single field display and field by field advance.

These authors emphasize the importance of the thermal properties of the material under investigation in relation to specimen thickness. For example, when the back surface (i.e. opposite the heated face) is being studied, the time taken after heating for the surface to attain half its ultimate temperature is

$$t_{\frac{1}{2}} = \frac{1.38 L^2}{\pi^2 \kappa} \tag{16}$$

where L is the thickness and $\kappa = K/C\rho$, ρ being the density, C the specific heat and K the thermal conductivity. Also of interest is the initial θ_f temperature generated at the heated surface, which may be a limiting factor in polymer-based solids. Table 5 gives values of $t_{\frac{1}{2}}$ and θ_f for a number of materials, together with the depth of the heated layer attaining θ_f. From this it can be seen that non-metals with their relatively low thermal conductivities are favoured by their long rise times compared to metals.

Results obtained so far show that observations of the face opposite the

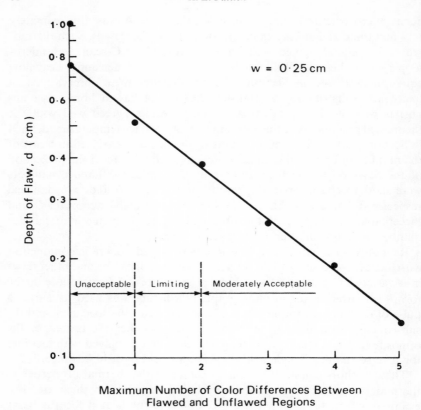

FIG. 13. Experimental number of colour differences in liquid crystal detectors between flawed and unflawed regions for flaws of the same width w, as a function of flaw depth.

heated surface (the so-called two-sided transmission technique) can detect deeper defects than the single-sided examination, but for defects close to the surface the dimension and depth resolution obtainable with single-sided transmission is superior. These methods have been successfully applied to the following:

(i) Detection of defects in CFRP aerofoil sections. Figure 14a–c shows three thermograms of the dark contrast due to excess resin and metal wire inserts, both by transmission and single-sided examination. The background structure due to the woven fibre is noteworthy.

(ii) Thickness variations in CFRP laminate sheets. Figure 15 compares

TABLE 5
Thermal Rise Times and Temperatures for Materials of Interest

Material	A (s)	B (K)	C (mm)
Mild steel	1.1	9.6	0.30
Stainless steel	2	13.0	0.22
Copper	0.13	3.4	0.87
Aluminium	0.15	5.2	0.80
Nickel	0.61	6.5	0.39
Titanium	1.8	21.0	0.23
Lead	0.61	16.9	0.39
Uranium	1.2	16.0	0.28
Brass	0.44	6.7	0.47
Zircaloy	2.3	20.4	0.27
Water	100	77.4	0.03
Resin	160	184.2	0.02
Glass	34	111.7	0.05
Porcelain	32	74.2	0.05
Concrete	26	87.9	0.06
CFRP ($\perp$ to fibres)	33	98.7	0.05
CFRP ($\parallel$ to fibres)	3.8	33.3	0.16
GRP ($\perp$ to fibres)	107	146.6	0.03
GRP ($\parallel$ to fibres)	82	128.2	0.03
Air	0.42	26 460	0.47

A: $t_{\frac{1}{2}}$ for sheets 1 cm thick
B: Maximum front face temperature θ_f with 6 ms flash depositing 1 J cm^{-2}
C: Thickness of material heated to θ_f

the results of single-sided thermography with those of low-voltage high definition radiography.

(iii) Debonding between aluminium honeycomb and CFRP skin. Figure 16 illustrates the opposite contrast obtained by thermal transmission and single-sided examination.
(iv) Bonding defects in plasma-sprayed surface coatings (Fig. 17).
(v) Location and sizing of holes in plastics (Fig. 18).

3.4.2. Molecular and fibre orientation

Most polymeric materials, whether filled or unfilled, possess a certain degree of anisotropy due to local alignment either of long-chain molecules or of the reinforcing fibres, which may lead to premature failure if a component is subjected to tensile stress in the direction normal to the direction of preferred orientation. The problem is particularly acute in

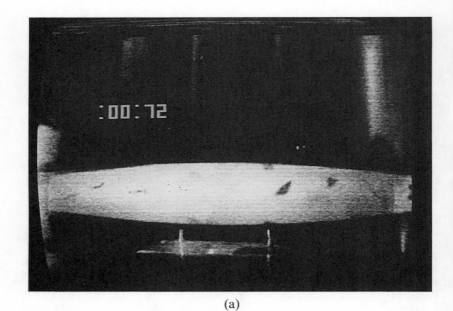

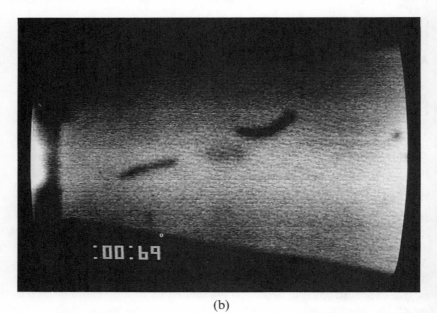

FIG. 14. (a) Excess resin and metal wire inserts in an 80 cm long CFRP aerofoil section shown as dark areas by thermal transmission; (b) similar inserts 36 mm long shown in detail. Note the woven fibre structure.

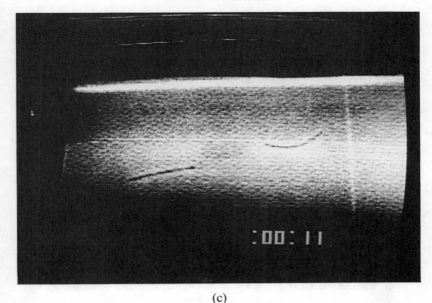

(c)

FIG. 14—contd. (c) Similar area viewed by single-sided technique.

injection mouldings, where the flow pattern often results in anisotropy which varies from point to point in the specimen, and with the increasing emphasis on this process in the manufacture of load-bearing components there is clearly an urgent need for a rapid non-destructive assessment of orientation for quality control.

Traditionally orientation in polymers has been detected by optical birefringence, mainly in connection with molecular chain alignment. This is in any case limited to transparent solids, and even in those does not always yield unambiguous results. In opaque materials the most direct way of observing mechanical anisotropy is by the use of ultrasonic shear wave velocity measurements [20], but this requires the use of contact probes and possibly some prior preparation of the specimen surface. It is therefore useful to supplement such measurements by the oldest and simplest method of characterizing anisotropy, making use of the variation with direction (in the surface) of thermal conductivity [16] in order to visualize the isothermals around a point source of heat. This technique was used by de Senarmont [21], who coated crystals with a layer of wax and touched the surface with a heated needle: the shape of the melted area

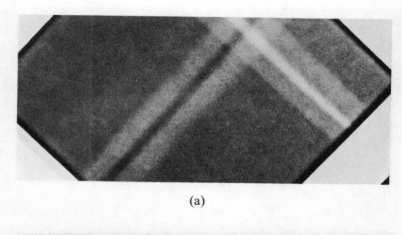

(a)

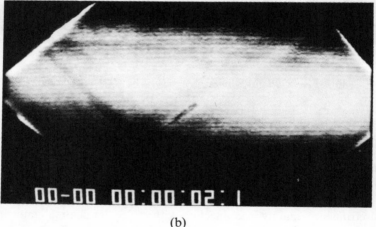

(b)

FIG. 15. Thickness variations in a sheet of CFRP laminate viewed by (a) low voltage high definition radiography; (b) single-sided thermography.

was in general an ellipse of which the ratio of lengths of the principal axes was proportional to the square root of the ratio of thermal conductivities. Modern thermal imaging of course renders possible the remote sensing of such temperature fields in a particularly convenient and accurate way. Liquid crystals were used as detectors in observations of this kind on oriented polymers [22].

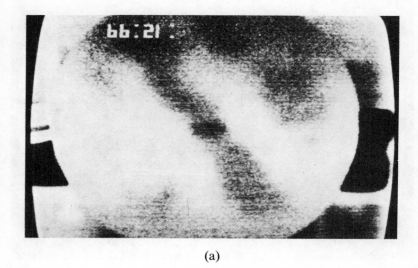

(a)

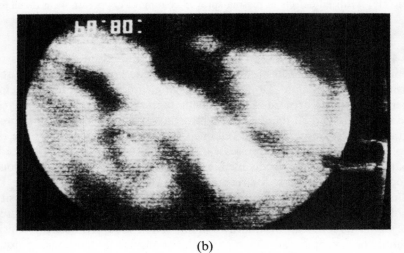

(b)

FIG. 16. 50 cm aluminium honeycomb with CFRP skin viewed by (a) thermal transmission; (b) single-sided examination. The poorly bonded areas show opposite contrast in the two cases.

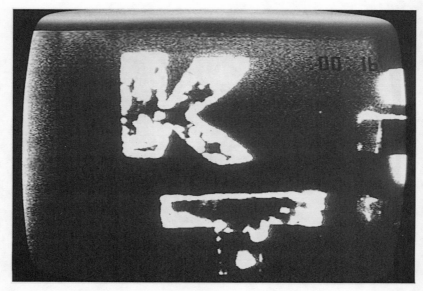

Fig. 17. Artificial defects in plasma-sprayed coating of copper on 4 cm diameter disc of mild steel.

Berrie et al. [8] observed the temperature contours on the surfaces of polymers and composites following the application of a point heat source using a Thermovision 780 infra-red imaging camera. The transient temperature field was recorded on magnetic tape and subsequently replayed to enable individual images, at 2 s into each cooling cycle, to be photographed. The output may be obtained on screen either in grey-scale or colour-coded form: the grey-scale display may be processed to isolate a particular isothermal for detailed inspection of its shape, while the colour coding is more suitable for recording the whole distribution. Figure 19 illustrates the range of possibilities for a single material, high density polyethylene. 19a and b show the pattern resulting from the heat source (2 s after its removal) on undrawn compression-moulded material which should be isotropic, and 19a in particular confirms that the single contour possesses circular symmetry to within the accuracy of the display. 19c and d, on the other hand, are photographs of the field obtained from the same material after uniaxial extension to a draw ratio of 12. The contours are elliptical with a ratio of lengths of major and minor axes of 4.9. This would correspond to a ratio of principal thermal conductivities in the

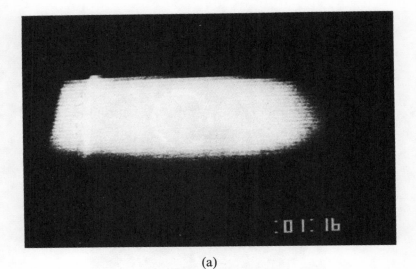

Fig. 18. Contrast in single–sided testing due to 1 mm holes drilled at various depths parallel to the surface in a block of thick painted perspex (a) 1 mm deep after 1.16 s; (b) 2 mm deep after 16.08 s.

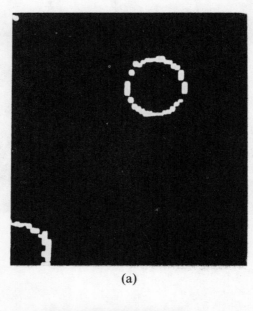

FIG. 19. Isothermal patterns of locally heated polymeric materials. Axis of draw horizontal. High density polyethylene (HDPE): (a) undrawn, single contour display; (b) as (a), temperature distribution axial ratio $b/a = 1$.

(c)

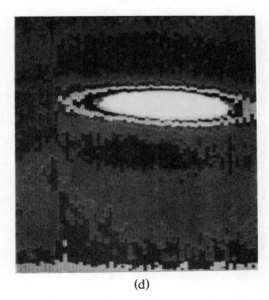

(d)

FIG. 19—*contd.* (c) draw ratio 12, single contour display; (d) as (c), temperature distribution axial ratio $b/a = 4.9$.

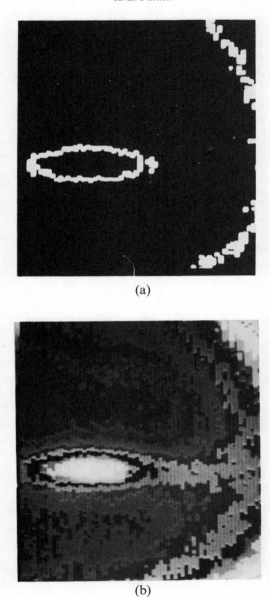

FIG. 20. Isothermal patterns of locally heated carbon fibre composite (fibre axis horizontal): (a) single contour; (b) temperature distribution. Axial ratio $b/a = 3.6$.

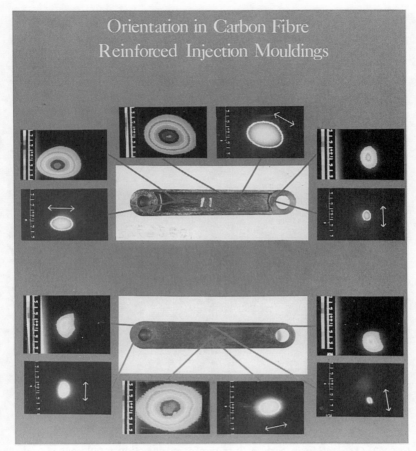

FIG. 21. Orientation in carbon fibre-reinforced injection mouldings (bicycle pedal crank). Local temperature distributions and single isothermal display. Double headed arrows indicate presumed preferred orientation direction. (a) Upper surface of component; (b) sectioned face.

surface of 24, of the same order as known conductivity ratios for drawn crystalline polyethylene.

The method is particularly well adapted to carbon fibre-reinforced polymers on account of the large differences in thermal conductivity between fibre and matrix, as well as the intrinsic thermal anisotropy of the carbon itself. Figure 20 demonstrates the effect on a conventional long fibre-reinforced epoxy resin composite, where the major/minor axis ratio

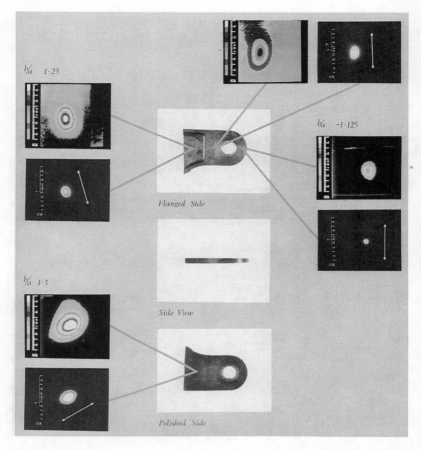

FIG. 22. Orientation in carbon fibre-reinforced injection mouldings. Local temperature distributions and single isothermal display. Double headed arrows indicate presumed preferred orientation directions; b/a values, approximate ratio of major to minor axes. (a) Upper surface of component; (b) sectioned face.

is ~ 3.6, corresponding to a derived principal conductivity ratio of 12.9.

The most important practical application of the technique is to the newer injection-moulded short fibre-reinforced thermoplastics, in which correct fibre orientation in the outer tensile load-bearing regions of engineering components is important. Figure 21 is an example of the transient local temperature distributions at various points on the surface of a typical article, a bicycle crank with an I-beam cross-section moulded in CFR-nylon 6,6. The variation in preferred fibre orientation direction is marked; in addition the shape of the temperature contours often departs considerably from the ideal ellipse, indicating the presence of local inhomogeneities such as surface voids or cracks. Figure 22 shows similar results on a second type of moulding.

It is clear that experiments of this type offer many possibilities for development into fully automated tests for non–destructive testing in production. A suitable system of mechanical scanning, or possibly simultaneous application of a number of point sources, would furnish information regarding the variability of local mechanical properties more quickly than any other method currently in use. Nor are applications limited to surface defects: through-thickness measurements of heat flow should be able readily to detect anomalies due to, for instance, out-of-plane fibre orientation. It should, however, be emphasized that no one technique can encompass all types of flaw in a given material, and it will usually be essential to monitor the results of two independent methods. A combination of thermographic and ultrasonic procedures [20] appears particularly promising for short fibre-reinforced injection mouldings.

ACKNOWLEDGEMENTS

I wish to thank authors cited for permission to use their results in this review. I am particularly grateful to Dr W. Reynolds and his colleagues of the NDT Centre, AERE Harwell, for the illustrations of their work, Dr J. G. Rider and Mr M. F. Markham of the Physics Department, University of Surrey, for helpful discussion of the ultrasonic applications, and Mr M. Rudman for assistance with the figures and photographs. Finally, I thank Mrs Sheila Rudman for drawing the diagrams.

REFERENCES

1. P. V. McLaughlin, E. V. McAssey and R. C. Dietrich, Non-destructive examination of fibre composite structures by thermal field techniques, *NDT International*, April 1980, **13**(2), 56–62.
2. K. L. Reifsnider and R. S. Williams, Determination of fatigue-related heat emission in composite materials, *Experimental Mechanics*, December 1974, **14**(12), 479–485.

3. J. J. Nevadunsky, J. J. Lucas and M. J. Salkind, Early fatigue damage detection in composite materials, *Journal of Composite Materials*, October 1975, **9**(4), 394–408.
4. E. G. Henneke and T. S. Jones, Detection of damage in composite materials by vibrothermography, *ASTM Special Technical Publication 696*, December 1979, 83–95; R. B. Pipes (ed.), *Proc. Symp. on NDE and Flaw Criticality for Composite Materials*, Philadelphia, 10–11 October 1978.
5. C. J. Pye and R. D. Adams, Detection of damage in FRP using thermal fields generated during resonant vibration, *NDT International*, June 1981, **14**(3), 111–118.
6. J. M. B. Webber, Principles of SPATE (Stress Pattern Analysis by Thermal Emission) technique for full field stress analysis, *1st International Conference on Stress Analysis by Thermoelastic Techniques*, London, November 1984.
7. W. N. Reynolds and G. M. Wells, Video-compatible thermography, *British Journal of Non-Destructive Testing*, January 1984, **26**(1), 40–44.
8. M. A. Berrie, K. E. Puttick, J. G. Rider, M. Rudman and R. D. Whitehead, Thermal probe analysis of orientation in polymers and composites, *Plastics and Rubber Processing and Applications*, June 1981, **1**(2), 129–131.
9. D. P. Almond, P. M. Patel and H. Reiter, The potential value of photothermal imaging for the testing of plasma sprayed coatings, *Journal de Physique Colloque*, October 1983, **44**(C–6), 491–495.
10. D. W. Wilson and J. A. Charles, Thermographic detection of adhesive-bond and interlaminar flaws in composites, *Experimental Mechanics*, July 1981, **21**(7), 276–280.
11. V. Vavilov, Infra-red non-destructive testing of bonded structures: aspects of theory and practice, *British Journal of Non-Destructive Testing*, July 1980, **22**(4), 175–183.
12. C. M. Sayers, Detectability of defects by thermal non-destructive testing, *British Journal of Non-Destructive Testing*, January 1984, **26**(1), 28–33.
13. J. A. Charles and D. W. Wilson, A model of passive thermal non-destructive evaluation of composite laminates, *Polymer Composites*, July 1981, **2**(3), 105–111.
14. J. H. Williams, S. H. Mansouri and S. S. Lee, One-dimensional analysis of thermal non-destructive detection of delamination and inclusion flaws, *British Journal of Non-Destructive Testing*, May 1980, **22**(3), 113–118.
15. S. F. Burch, J. T. Burton and S. J. Cocking, Detection of defects by transient thermography: a comparison of predictions from two computer codes with experimental results, *British Journal of Non-Destructive Testing*, January 1984, **26**(1), 36–39.
16. H. S. Carslaw and J. C. Jaeger, *Conduction of Heat in Solids*, 2nd edition, Oxford University Press, Oxford, 1959, ISBN-0-19-853303-9.
17. L. J. Broutman, T. Kobayashi and D. Carillo, Determination of fracture sites in composite materials with liquid crystals, *Journal of Composite Materials*, October 1969, **3**(4), 702–704.
18. J. A. Charles, Liquid crystals for flaw detection in composites, *ASTM Special Technical Publication 696*, R. B. Pipes (ed.), December 1979, 72–82.
19. S. S. Lee and J. H. Williams, Thermal non-destructive testing of fibre glass laminate containing simulated flaws orthogonal to the surface using liquid

crystals, *British Journal of Non-Destructive Testing*, March 1982, **24**(2), 76–81.
20. J. D. Aindow, M. F. Markham, K. E. Puttick, J. G. Rider and M. R. Rudman, Fibre orientation detection in injection-moulded carbon fibre reinforced components by thermography and ultrasonics, *NDT International*, February 1986, **19**(1), 24–29.
21. H. de Senarmont, Mémoires sur la conductibilité des substances cristalisées pour la chaleur, *Comptes Rendus*, 1847, **21**, 459/707/829.
22. H. G. Kilian and M. Pietralla, Anisotropy of thermal diffusivity of uniaxial stretched polyethylenes, *Polymer*, June 1978, **19**(6), 664–672.

Chapter 4

Optical Methods

COLIN A. WALKER and JAMES MCKELVIE
Department of Mechanics of Materials, University of Strathclyde, Glasgow, UK

4.1. THE AIMS OF NDT IN RELATION TO FIBRE-REINFORCED COMPOSITES

4.1.1. Definition
For the purposes of this section, we will define NDT as an evaluation of the physical parameters of a material or structure by non-destructive means. The implication of this definition is that certain topics will be covered which are more usually considered under the heading of metrology or inspection.

4.1.2. Approach
In subsequent sections each technique will be discussed separately, along with examples of its application and an evaluation of its overall potential usefulness. In view of the rapidly changing state of affairs, a separate section will deal with image analysis, and the interfacing of the various techniques with computer systems, to create automated or semi-automated NDT and inspection systems.

4.1.3. The particular problems posed by fibre-reinforced materials
Before considering specific aspects of the problems that arise in the inspection of fibre-reinforced materials, we should first of all take note of the diversity of structures involved. There are, at the simplest level, sample coupons of material which are being used to evaluate basic material properties; we then proceed through sub-assemblies to complete struc-

TABLE 1

Application / NDT task	Material properties	Fracture studies	Model evaluation	Whole structure monitoring
Dimensional control			✓	✓
Load–deformation curves	✓	✓		
Flaw detection		✓	✓	✓

tures, which may involve core materials that range from foam plastics up to expanded hexagonal cellular structures fabricated from metal or fibre-reinforced plastic; and these cores, in addition, will be overcoated with a multilayer laminate. Faced with this diversity, one can discern not a single task of NDT, but a series of differing requirements, each one a component in the overall scheme of structural evaluation, and each of these requirements will demand its own solution (Table 1).

Equally, we should recall the comparative youth of fibre reinforcement technology, which is evidenced by the need not only to measure the physical characteristics of structures during loading, including the identification of flaws, but also to be able to quantify the effects of flaws and dimensional changes upon the overall structure; to this extent, the task of NDT in fibre-reinforced structures differs from that in metals since there is a more complete understanding of the behaviour of the basic metallic materials.

It should, perhaps, be pointed out that problems exist, of a specifically technique-related nature, that revolve around the fact that fibre-reinforced materials are (a) by-and-large opaque, and (b) by design non-isotropic. We are, therefore, often in a situation with optical techniques where we are forced to use surface measurements to assess the nature of events occurring some distance below the surface, and it is this turn of events which can lead to the use of a combination of techniques; for instance, ultrasonic or radiographic methods may be used to characterise subsurface defects, while optical techniques may be used to show their structural significance.

In view, then, of the complexity of the task in hand, it is possible to assemble an application matrix in which a list of NDT objectives, on the one hand, is related to a list of optical techniques (Table 2).

Those objectives which may be addressed by a particular technique are indicated; a separate annotation is used where the optical technique would be supplemented by additional NDT measurements.

TABLE 2

Nondestructive testing objective	Candidate optical technique									
	Triangulation	Shadow moiré	Projected fringes	Moiré interferometry	Holographic interferometry	Speckle interferometry	Speckle photography	Shearography	Photoelasticity	Fibre optics
Dimensions	✓									
Step height	✓	✓	✓							
Area		✓	✓							
Gap width										
Shape (silhouette)										
Edge profile										
Surface profile	✓	✓	✓		✓					
Fracture toughness				✓	✓	✓	✓	✓	✓	
Fatigue crack growth				✓	✓	✓	✓	✓	✓	✓
Load distribution				✓			✓		✓	
Deformation				✓	✓	✓	✓	✓		
Strain field				✓					✓	
Comparison (objects)	✓	✓	✓							

} *These headings may be treated under 'Visual inspection'.*

4.1.4. General discussion of optical NDT techniques

The appeal of optical techniques is fairly easily understood from a historical perspective, since, in some shape or form, optics has been used for alignment and gauging since time immemorial; there is, too, the reputation that optical methods have for extreme precision; and finally, there is the point that most optical methods generate results which are pictorial, often elegant and, at least from a qualitative point of view, easily understood.

To a large extent, optical methods have been changed almost beyond recognition in recent years (say, post-1960) by the development of new light sources, detectors and electronic systems. A well-known example here will, perhaps, suffice to demonstrate the point. The principles of holography as we know them today could easily have been enunciated 100 years ago, and are, indeed, implicit in Abbé's theory of the microscope: certainly, in Gabor's original work on holography in 1948, the ideas of image recording are well founded, yet it was not until the 1960s, after the development of (a) the laser, (b) a wider appreciation of the theory of coherence, and (c) suitable films, that the first successful holograms were made.

The situation now exists where lasers are reliable, cheap, and available in a variety of wavelengths and powers; their use is accepted, and safety regulations have been refined, and enforced, to the point at which even high-power lasers can now be used with relative safety. Likewise, the photomultiplier has been replaced for many applications by a range of solid state devices, including single photodiodes, and photodiode arrays, both linear and biaxial; these latter image-plane arrays currently range up to 512×512 unit diodes, and so are in direct competition, as far as resolution is concerned, with vidicon tubes for imaging applications.

Finally, this discussion should underline the contribution made by electronics to the application of optical methods; this ranges from simple points like the provision of power supplies for lasers and detectors, through signal generators to full-scale data processing and, in recent years, image processing systems. While one may recall the ready appeal of an image as the output of an NDT system, the useful data are often obtained by an automatic or semi-automatic quantification of the salient features of the image (e.g., as in Figs 5a and 5b).

4.2. VISUAL INSPECTION

4.2.1. Imaging systems

The techniques considered under this heading may be considered as

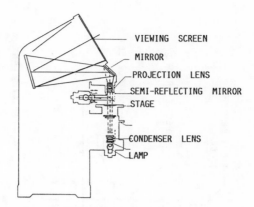

FIG. 1. Profile projector.

relatively conventional, since imaging optics have long been used for metrological purposes. They remain in widespread use for dimensional measurement and for the inspection of shape and linearity.

Typical of such systems is the profile projector, which uses a collimated beam of light and an optical system to throw onto a screen a magnified image of the outline of the object (Fig. 1). This system finds a wide range of uses, particularly in the checking of complex parts against a master drawing on the projection screen. A resolution of about 2.5 μm is routinely achieved for small components. The almost universal availability of the profile projector, and the ready acceptance of its use, makes it one of the most used and useful of optical devices in the metrology field. Plainly, the application of the technique to larger structures depends upon specialised optical systems, with proportionally lower levels of accuracy obtainable.

4.2.2. Three-dimensional shape measurement

A number of devices have been developed for measuring absolute three-dimensional shapes; these devices adopt a number of different working principles, and for this reason some will be considered in this section, while others will appear in a later section on laser scanning.

The triangulation approach was used by Ford [1] to measure variations in depth over a small range (Fig. 2) of 0.25 mm. The two photodetectors measure light scattered from the surface, which is illuminated by two overlapping spots derived from the LED/prism arrangement. The device shows a somewhat disappointing dynamic range, since its resolution

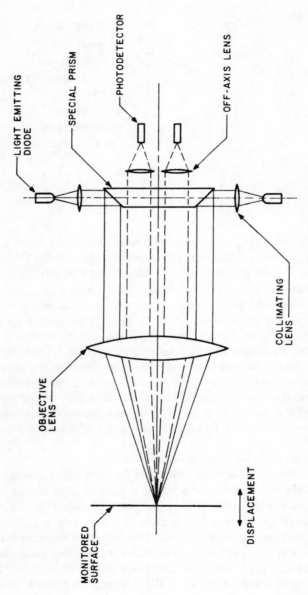

Fig. 2. Depth measurement using a triangulation principle.

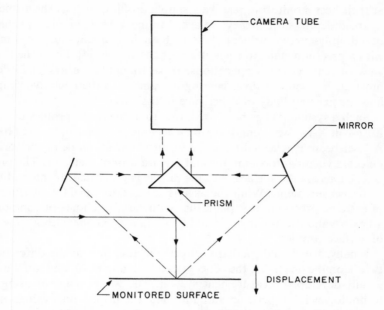

Fig. 3. Triangulation device for measuring tyre tread depth.

(0.025 mm) is only 1/10 that of the linear range. From that point of view this technique must be considered of limited usefulness.

Another technique which has, at first sight, an almost charming simplicity (Fig. 3) was developed for measuring tyre tread depths, and as such must be considered to be of proven utility [2]. The laser beam is directed normally onto the surface, and the scattered light is imaged onto a vidicon camera tube. The distance between the spots on the tube is directly related to the distance of the test surface from a reference plane. The sensitivity of this system is such that it can locate the test surface accurately to 0.0025 mm over a range of 50 mm; the sensitivity (and the range) can be altered to suit the application in hand by altering the interrogation angle.

Similar systems, using triangulation principles, have been developed for measuring the profile of steel slab and sheet during rolling [3]; these methods have the advantage of conceptual simplicity; as we have seen, their sensitivity is consistent with the requirements of precision structural design, and they are easily interfaced with computers. They are probably at their best monitoring relatively simple profiles, such as sheet thickness or tread depth, since if they are asked to measure a complete profile (say of a doubly-curved surface), then the basic instrument must be combined with a precision stage to track it across the surface [4]. From the point of view of real systems applications, these triangulation systems offer the most at this point in time, bearing in mind both their relative simplicity and the existing body of experience in their use.

Such a system [5] has been used, too, to measure the profiles of mirrors for use at X-ray wavelengths with an accuracy of one nanometre (10^{-9} m). A laser beam is reflected from the surface, and the slope of the surface is detected via an autocollimator and quadrant photodetector. The probe is tracked across the surface, and the profile is generated by integrating the slope measurements. While the technique was developed for measurements of extreme precision, it is plain that with suitable choice of components, a flexible and precise device could be designed to cope with a wide range of surface profiles.

Finally, we should, perhaps, direct our attention to the difference between, on the one hand, the concept of a 'gauging head' and, on the other, a fully-developed measurement system. The whole point of using these technologies is to speed up and automate the inspection/evaluation process, and it is no small matter to conceive of, design, and bring to fruition, a total package based upon these ideas. As an approximate rule of thumb, one may estimate the cost of the optical part of the overall system as being between 1 and 5% of the total, the remainder being mechanical scanning components, computer hardware, and software development. In consequence, a complete, automated system can cost as much (or as little, depending on one's point of view) as \$250 000 (mid-1984).

4.3. LASER SCANNING

4.3.1. Technology review

For problems involving the quality of a surface, an approach that is often adopted is the use of a surface scanning system, in which a spot of light is scanned over the surface in a raster. Flaws in the surface result in changes in the reflected light intensity, enabling them to be quantified. By and large, this technology has been applied to strip materials where the speeds of production are such that visual inspection is out of the question. If we refer, now, to composite materials, the volume of production is scarcely at a level to justify the use of such techniques.

4.3.2. Laser scanning and slope measurement

However, a laser scanning, surface contouring method has been developed by Himmel [6], using a pulsed laser beam and a sensitive system for measuring the slope of the surface. The problem with slope sensing is that any error is propagated throughout subsequent measurements, and thus frequent recalibration on a reference surface was felt to be necessary. Notwithstanding, a system accuracy of ± 0.05 mm was claimed over a contour slope range of 75°. Plainly, the scanning approach has much to recommend it, since it does not involve the mechanical scanning of object or optics, and it may be seen as intermediate between triangulation systems and the whole-field systems discussed in a later section; the system is easily adaptable to changing geometry and is by its very nature a computer-based system, since the output appears as a time-series of fluctuations in the voltage from a photodiode.

4.4. MOIRÉ FRINGE METHODS

4.4.1. Introduction

In general, the term moiré denotes the use of a regular pattern in the form of a series of parallel, or crossed, lines. These patterns may be material and visible, as in the coarse gratings used in demonstrations of the moiré effect, or they can be patterns of light and dark projected in space; almost all variations between these extremes have been used at some stage or other for testing purposes. It is convenient to divide the family of moiré techniques in two, as follows: (a) those where a grating is applied to the surface for the measurement of in-plane strain, and (b) those methods which rely upon a grating projected optically onto the surface; these projected fringes are used in the main for the analysis of surface form, and it is these which we will consider first.

4.4.2. Measurement of shape and out-of-plane deformation

4.4.2.1. *Shadow moiré*

Conceptually, the simplest technique [7] for the measurement of shape is shadow moiré, in which a grating is held just clear of the surface and illuminated obliquely; from a viewpoint normal to the surface, the lines of the grating combine with the lines projected on the surface to create a series of surface contours, whose spacing may be expressed by:

$$h = S_0 \cot \varphi$$

where h is the spacing of the contours, φ is the angle of illumination, and S_0 is the grating spacing.

From a practical point of view, shadow moiré is an attractive proposition, since a useful system can be set up with a minimum of apparatus—for example, using a woven nylon screen with a 4 thread/mm weave (as is readily available), and using an illumination angle of 45°, we have a contour interval of 0.25 mm. Where the object is of such a size that the illumination angle may vary, steps to accommodate this change may be required. However, when the system is being used in its NDT role, it will usually be operating as a comparator, i.e., comparing the surface form of a component under different load levels, or comparing two components to ensure that they are the same shape. In this configuration, the geometry is normally well defined and allowance can easily be made.

A further problem may arise from the shadowing of vital detail by the offset illumination, but this can be avoided by arranging for the illumination to be placed symmetrically about the point of view.

One of the most attractive features of shadow moiré is the ease with which the sensitivity can be changed—either by using a different screen, or by altering the illumination angle. The range of contour interval extends from 1/10 mm up to 5 mm. This technology has been used for the analysis of complex shapes such as the human body, and gear profiles.

One should add that the amount of data contained in one contour image is large. To make real sense in a routine manner, interface with a computer system is almost essential; even so, the quality of the fringes is excellent, presuming incoherent illumination, so that the image-processing task can be relatively straightforward (see Section 5).

The limitation of shadow moiré's capability tends to arise from the requirement for the grid to be close to the surface under scrutiny: with increasing depth of the surface below the grid, the contrast of the fringes is reduced.

4.4.2.2. *Fringe projection*

The limitation which has just been discussed with reference to shadow moiré, on the depth of specimen which can be handled, is much less bothersome in systems which rely upon the projection of a regular series of lines onto the specimen surface (Fig. 4). The specimen, with its series of lines, is recorded on film with a master grating held just in front of the film plane. The film records the moiré beat between the specimen grid and the master. Since the photography may be carried out from the same direction as the fringe projection, shadowing need present no problem; likewise,

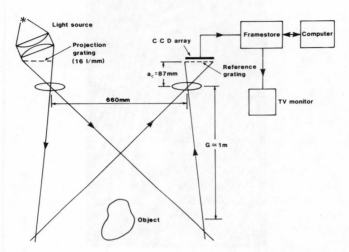

FIG. 4. Fringe projection contouring system (courtesy of Mr G. Reid, NEL, East Kilbride).

provided the fringe projector and the camera lens have sufficient depth of field, the depth range of the system is greatly increased compared with shadow moiré, at a contour interval that covers the same approximate range upward from 0.1 mm [8].

Results, typical of those obtained with this technique, are shown in Fig. 5. [9]. These results were produced by a system in which the projected master grating is moved by $\pm$ half a pitch at right angles to the grid lines. At each point in the sampled image the relative phase of the signal can be calculated from the three intensity readings, and by this means the shape can be reconstituted free from the effects of noise, which tends to cause problems in any automatic inspection system.

A number of modifications have been proposed to the basic method with the requirements of an inspection technique in mind. For example, if a standard object exists with which the test object is to be compared, then the standard grating that is placed close to the camera focal plane can be generated by photographing the standard object with the fringe pattern projected onto it. Any differences in shape between the standard component and a production item (or between the same component before and after loading or service usage) will then appear as a series of moiré fringes; this situation is easier to analyse than one in which an operator, or a computer, is being asked to carry out the shape subtraction.

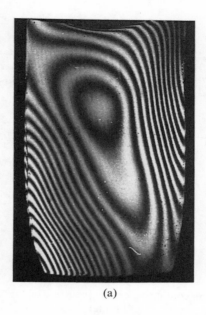

(a)

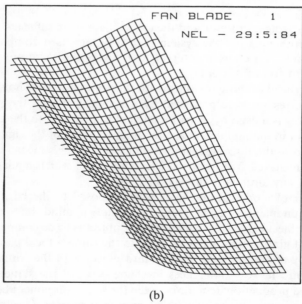

(b)

FIG. 5. (a) Projected moiré fringe pattern. The master grating is clearly visible. The shape contours shown in (b) were computed using two additional patterns generated by moving the master grating by $\pm$ half a grating space. (b) Contour shape derived from the projected moiré photograph of (a).

Since the creation of the perfect standard component may give rise to major difficulties, it has been suggested that the master grating could be generated by computer graphics methods, knowing the desired geometry of the component and the illumination system. Another approach, demonstrated by Nordbryn, [10] is to view the projected fringe pattern with a CCD camera, and to arrange the density of the fringes so that a moiré beat is seen between the fringes and the regular array of the camera elements. By this means, a variable contour interval over a range between 1 mm and 10 mm can be made available, without the specific need for a master grating.

4.4.2.3. *Hardware implications*
At this stage, it is useful to look at some of the actual hardware components used in moiré topography. Shadow moiré has the simplest requirements, which may run merely to a screen of a suitable mesh size stretched over a frame, a compact-filament lamp and a 35 mm camera. If parallel illumination over a large area is desired, an inexpensive Fresnel lens can be used for apertures up to 30 cm in diameter.

A large number of devices have been used for fringe projection, varying from bar-and-space gratings in projectors to sophisticated interferometers for projecting interference fringes. It is as well to note that the demands placed upon the optics are much higher in fringe projection, since the moiré beat is formed at the focal plane of the camera rather than in the object space. Accordingly, for accurate work on objects of a useful size (say 50 mm or more across), a careful selection of the optical components, both for fringe projection and for viewing, is indicated, to ensure that their performance is compatible with the inspection performance desired.

4.4.3. Measurement of in-plane strain
4.4.3.1. *Definition of the problem*
Compared with the measurement of shape, the measurement of in-plane strain is a comparatively difficult task, since the quantities to be measured are so small. One may get some idea of the task in hand by examining likely levels of deformation; typically, a fibre-reinforced material will reach a strain of 5% before fracture, i.e., it stretches by one part in twenty of its original length. If we imagine, now, that we are trying to evaluate the effect of loading upon the detailed structure of a specimen, we may wish to view changes in the induced in-plane strain over distances as small as 0.1–0.2 mm. This implies that the measurement system must be capable of detecting changes which are at *most* twenty times smaller ($<5\%$), i.e.,

5–10 μm. For a useful system, then, it is plain that one should be able to detect deformations as small as 1 μm over distances of a fraction of a millimetre. This combination of extreme accuracy and short gauge length is by no means universally necessary—for example, one can measure the deformation in a 150 mm long plain specimen with equipment little more complicated than a ruler; equally, one can state that problems requiring extreme specifications are numerous in composite NDT.

4.4.3.2. *Bar-and-space gratings*

The accuracy of any moiré system depends upon the spacing of the grating in use. The first gratings used for moiré were photographic bar-and-space structures. The limiting density for such gratings is about 40 line/mm, implying that the resulting deformation contour interval will be 1/40 mm. The main use for such gratings has been in the visualisation of loading

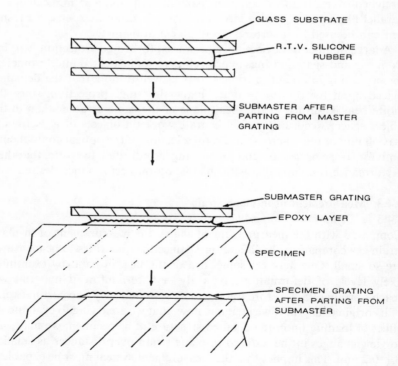

FIG. 6. Replication process for the phase grating system.

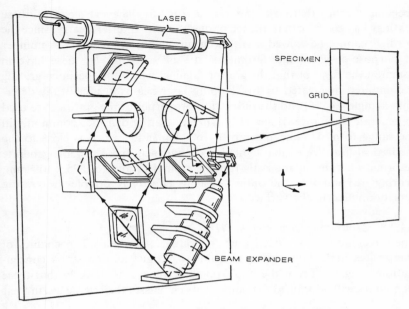

Fig. 7. Schematic of portable moiré interferometer.

effects in deformable models of structures [11], or in evaluating large strains such as occur during the welding of metals [12].

One notable exception to the general rule is the camera modification devised by Burch and Forno [13], which allows gratings to be photographed at reasonable densities. The image-plane line density of 300 line/mm implies that a specimen with a grating applied at a density of 30 line/mm can be photographed at a demagnification of 10:1. In fact, this system is most often used in a double-exposure mode, with the load change between exposures giving the moiré effect. Despite its attractions as a compact method of recording gratings, it is still limited by its ability to handle specimen gratings of only modest density compared to those required for elastic in-plane strain measurement.

4.4.3.3. *Phase gratings*

In recent years it has become possible to create gratings which are much finer than the black-and-white type and are, at the same time, more efficient in their use of the available light. Such gratings modulate the phase rather than the amplitude of the light, and can be replicated onto specimens using thermosetting resins (Fig. 6). The existence of these gratings has given birth to the technique of moiré interferometry, since the gratings cannot be viewed by the conventional moiré method. A number of purpose-designed interferometers are in use (Fig. 7); these have in common the concept that the grating, against which the specimen grating is compared, is created by overlapping laser beams at a carefully determined angle [14–16] and the diffraction properties of the gratings are used to generate the moiré fringes (Fig. 8). The grating densities commonly in use range from 475 line/mm up to 1600 line/mm, which is close to the theoretical limit for visible light; in consequence, these systems generate deformation contour intervals ranging from 10^{-3} mm to 0.25×10^{-3} mm, bearing in mind that the optical systems effectively double the grating density applied to the surface.

4.4.3.4. *Moiré interferometry in NDT*
These techniques have been used to evaluate the detailed mechanics of composites under load, their reaction to fatigue and to stress concentrations (Fig. 9). From the NDT point of view, it is unlikely that these techniques will be applied to components en masse; their value lies in their

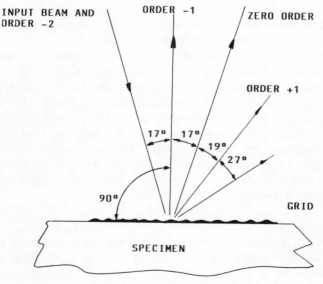

Fig. 8. Diffraction at a periodic grating.

ability to contribute real data to theoretical composite mechanics, and to assess the effect of defects upon the load-carrying capacity of structural components. A suitable level can then be set for defects to be detected by other NDT methods, consistent with the safe operation of the structure as a whole.

It will be seen from Fig. 9 that the fringe patterns which are recorded

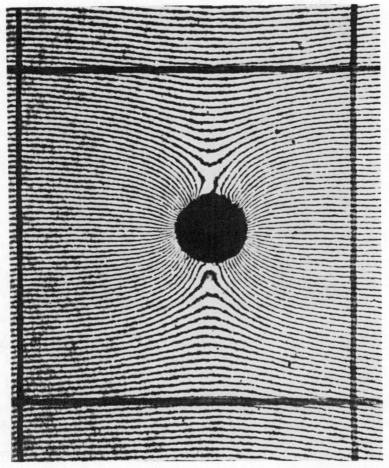

FIG. 9. Deformation contours (vertical component) on a fibre-reinforced coupon with stress concentration.

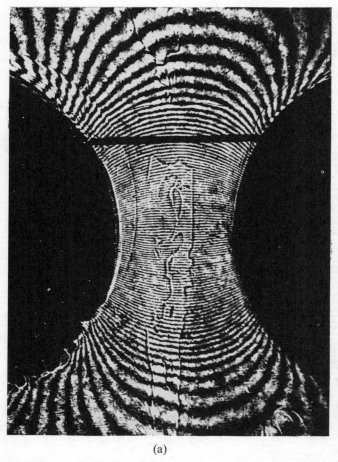

(a)

Fig. 10. Strain contours (b) derived from a deformation contour interferogram (a). The specimen is a square prism with hemicylindrical cutouts. The loading is vertical.

may show a high degree of complexity. Each loading condition requires three pictures to describe the strain field; there is, in consequence, a major data handling problem in prospect. Analysis of the interferograms is now often undertaken by interfacing digitising equipment with computers. The computer can then calculate the strain level at each desired point in the field, and output its results as a strain contour plot (Fig. 10).

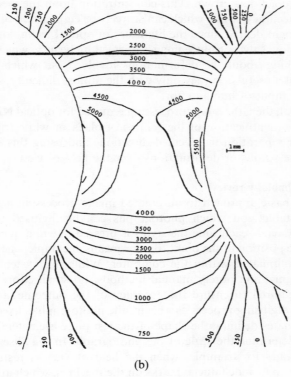

(b)

FIG. 10—contd.

4.5. IMAGE PROCESSING

Anyone who has more than a nodding acquaintance with the Apollo project that put a man on the Moon will be aware of the part played by the computer enhancement of images—how individual pictures were patched together to give a panoramic view, with minimal jumps in the brightness level across the borders. Miraculous as this seems (and despite the consequent impression that image manipulation is a relatively easy task), the image processing used for NDT purposes has a more mundane aim—the acquisition of data. There exists, also, a constraint that limits the use of certain of the more exciting approaches. The particular constraint is that it is the purpose of optical NDE to put numbers onto images—to specify how large are the strains, how far a panel is bulging from the true, how far a crack has grown. This concentration upon the measurements, and their accuracy, is one which can scarcely be emphasised often enough, or in sufficient degree, since the literature of optical technology is replete with techniques that produce images of startling beauty, which very often say something about the deformations involved but which can deliver numbers of a useful accuracy only with the utmost difficulty, and after an inordinate input of human effort.

In general, then, the image processing required for optical NDT revolves around the identification of the centres of black or white fringes, measuring their separation in specified directions and using this information to calculate strains or deformations over the field of view.

4.5.1. Technology review

The most basic approach to automated image processing is the manual digitising tablet, on which photographs can be digitised with a pen-follower system—each time the stylus is pressed down, a microcomputer records the position of the point. By this means, a complete field of fringes can be manipulated within the computer. While this system is a major improvement upon purely manual methods, no small effort of organisation is required to ensure the accuracy of the outcome; for instance, before the digitising process can begin, the photograph is fixed down and the coordinates defining the shape of the test piece are entered. This also serves to determine the scale of the photograph. In the absence of component detail—for example, when the field of view is restricted—it is mandatory to include fiducial marks on the specimen which appear on the final picture. For the fringe analysis itself, it is usual to adopt a system of giving each fringe a number, and of identifying this number before the fringe is digitised in a standard number of steps. This data is then used to construct a three-dimensional surface in space, using a cubic spline routine to fit polynomials to the experimental data. It is then a fairly simple matter for the computer to extract deformation data at any point, or along a chosen section; if strain values are required from a displacement field, the cubic splines are differentiated and the slope evaluated at each point to give the strain. With the strain component information in this form, a further calculation step will combine three deformation fields to give principal strains and their directions.

The semi-manual method of digitising and image analysis has been treated at some length, since it shows up the problems that one faces in constructing a fully automatic analysis system. Firstly, the photographic process, while often viewed as a major drawback in optical systems, has the advantages of permanence and data storage capability; video systems, while much more convenient, have a limited resolution and, save for special units with built-in shutters, are limited to a minimum exposure of 1/25 s; image motion may, therefore, give rise to problems. Whatever the origin of the data, the automatic system is faced with the problem of coping with optical noise in addition to the factors outlined above. A great deal of effort has gone into evaluating the effect of minor irregularities in fringes upon the overall accuracy. The fact remains, however, that in an unknown strain field which is varying rapidly, one can but assess the gauge length required for one's purpose, and then live with the consequent accuracy, bearing in mind that gauge length and accuracy bear an inverse relationship to each other (see Table 3).

TABLE 3

Relationship Between Gauge Length, Strain, Spatial Resolution and Accuracy

Fringe frequency (fringes/mm)	Strain level (microstrain)	Accuracy of strain measurement (microstrain)	Spatial resolution (gauge length, mm)
0.1	105	±5	10
1	1 053	±50	1
10	10 526	±500	0.1

It is scarcely surprising, then, that very little data exists about the actual performance of automatic fringe-analysis systems of a general nature; there are, however, a number of special-purpose analysis systems which have been successfully demonstrated, and these will now be discussed.

4.5.2. Automatic fringe analysis systems

In general (Fig. 11) systems comprise a number of modules which may be combined in various ways; it is normal practice, now, to retain the output of a video camera in a rapid access frame store which can be accessed, pixel by pixel, for processing. Typically, a frame store will hold a frame of video at a resolution of 512×512, i.e., there are 512 lines of video information with 512 discrete points on each line; at each point the image intensity is characterised by an 8-bit number i.e., to an accuracy of 1 part in 256 or $\sim 0.4\%$. The heart of the system is a medium-range mini-

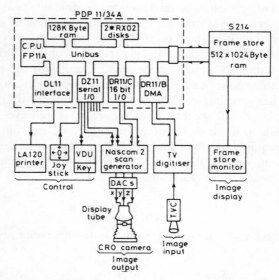

FIG. 11. Schematic of an automated image analysis system. (After D. Robinson, NPL, UK.)

computer, and the final information is output on a video display unit for immediate viewing, or on a high-quality printer for permanent record.

Typical problems include the quantification of specklegrams [7] (see Section 4.7.2.2), where the desired data are the fringe separation and orientation (Fig. 12). The approach is largely determined by the grainy nature of speckle fringes, and proceeds as follows:

(1) The direction of the fringes is determined by summing the intensity of the image along a vector which is rotated through 360°. The intensity is highest when the vector lies along a bright fringe.
(2) The whole image is reduced to a one-dimensional intensity distribution by summing all the intensities along the direction found in step 1. By this summation technique, the noise in the fringes is averaged out and therefore greatly reduced. A simple linear analysis now yields the peak-to-peak spacing for the one-dimensional intensity distribution.

This process can be completed in a few seconds (say ~ 5 s). For a complete analysis of the specklegram, it will be recalled that it has to be stepped to a new position and the analysis repeated. Despite their apparent complexity and cost, such systems can perform complete analyses of speckle-

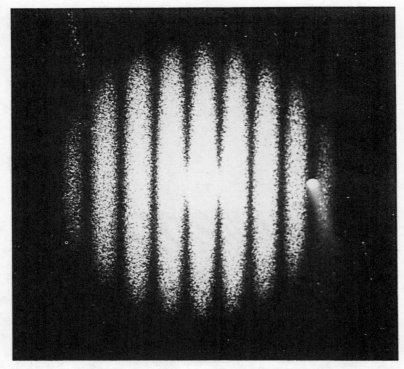

Fig. 12. Young's fringes arising from the interrogation of a typical speckle photograph. (Courtesy of Dr W. King, University of Strathclyde.)

grams quickly and accurately, and similar techniques can be applied to any pattern with linear or circular symmetry.

A second example is of particular interest in the NDT of composites, since the application is the detection of debonds in brazed panels; in many ways these are similar to composite panels in which the fibre-reinforced skins are bonded to a honeycomb core [17].

The panels were heated, and the debonds appeared as areas of closed ring fringes in the holographic interferogram (Section 4.6). Since one can arrange for the hologram to have few fringes other than the ring systems, these can be detected automatically by the property that a unit line vector scanned across the field will encounter the same number of fringes, irrespective of the orientation of the vector, when it is crossing a diameter of the circle. Points in the field which satisfy this criterion can be identified, and a decision can be taken whether to pass or fail the panel.

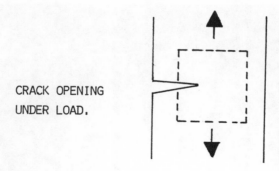

CRACK OPENING UNDER LOAD.

FIG. 13. Measurement of the J-integral fracture toughness parameter. Deformation contours around a crack tip. The J-integral is calculated along the indicated path.

The final example is taken from the analysis of moiré fringe patterns for the purpose of measuring the *J*-integral fracture mechanics parameter [18] (Fig. 13). While *J*, defined as

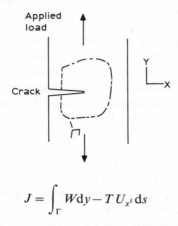

$$J = \int_{\Gamma} W \mathrm{d}y - T U_{x^i} \mathrm{d}s$$

where T is the traction vector normal to Γ, U is the displacement vector, and W is the strain-energy density, can be derived from a set of three complete frames of information, (x, y and 45° deformation fields) it is difficult, at high levels of strain, to extract sufficient data from the video camera due to the limitations on resolution and, having acquired it, to store it. The answer lies in recognising that only data around the integration contour Γ is important, so that the vidicon tube can be made to scan only areas close to Γ. By this means, resolution can be greatly improved, and the data storage problem reduced by a factor of 1000. Once the scan information is acquired, the value of *J* can be computed, knowing the material constants and the stress–strain curve.

4.5.3. Summary

It will be seen, then, that computer-assisted image processing is far from being a magical solution to quantitative picture analysis; a workable system requires a large helping of prior thought, an evaluation of the extent to which the particular problem in hand has its own peculiarities which can be used to ease the task, and, during the developmental phase, a steady aim to be kept on achieving the desired end result rather than on contemplating the ingenuity of the hardware. With these caveats in mind, however, it is plain that the development of these systems is a major advance in the application of optical methods to NDT.

4.6. HOLOGRAPHIC INTERFEROMETRY

4.6.1. Classical holography

4.6.1.1. *Introduction*

A vast amount of work has been published on the use of holography for engineering inspection. The attention the technique has received is entirely understandable, since it holds out the possibility of completely characterising the form of a surface (and by implication any changes in it) by doing nothing more than shining light on it. With such a tool, it is possible in principle to derive surface displacement, strains and stresses; to see the modes of vibration; to detect and quantify cracks; and to diagnose subsurface phenomena by virtue of their effect at the surface. Shape-comparison of two different objects is even possible in principle. In reality, however, the success of the technique has been quite dichotomous. Whereas for qualitative or semi-quantitative inspection for flaws, cracks, debonds, vibration analysis etc it has proved to be a relatively practical tool, its usefulness in quantitative stress and strain measurement has been very restricted.

4.6.1.2. *The principles*

To understand this contrast of behaviour requires an appreciation of the principles, which, very briefly, consist of capturing not only the information regarding the intensity of light reflected at each point of the surface (as in ordinary photography) but also its phase. This is done by illuminating the photographic plate with a 'reference beam' of a known, or at least reproducible, geometry (e.g. a collimated beam or one having a spherical wavefront) *simultaneously* with the light scattered from the object (the 'object beam'), the two beams being mutually coherent (see Fig. 14a). Interference effects between object and reference beams produce the hologram. By 'reconstructing' the hologram with a geometrically similar reference beam, there emerges from the hologram, due to diffraction effects, an ensemble of light rays which is indistinguishable from the original ensemble scattered by the object (see Fig. 14b). By superimposing this reconstructed ensemble corresponding to the object in some condition, 'A', onto a beam corresponding to the object in some other condition, 'B', there are created, due again to mutual interference, light and dark fringes which can be related mathematically to the change in shape. The superimposition is termed 'holographic interferometry'. Typically, one

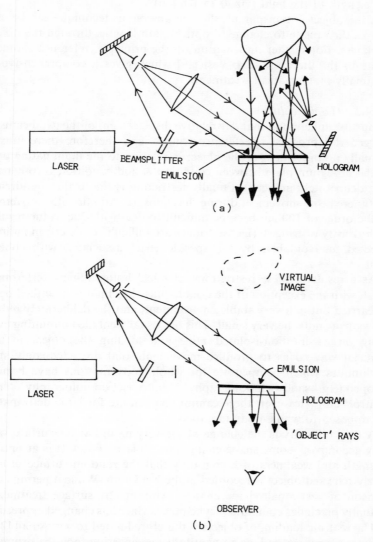

FIG. 14. (a) Formation of a hologram; (b) reconstruction of a hologram.

fringe interval corresponds to a displacement of between a half and one wavelength of the light (viz. 0.25–0.6 μm).

If the object is transparent, then an analogous technique can be used; fringes then relate to change of optical path lengths through the object's thickness. For further information on the principle, reference should be made to the literature, e.g. Vest [19] which gives a comprehensive but minimally mathematical treatment.

4.6.1.3. *The difficulties*

It must be understood that *any* relative displacement of the two beams will (in general) produce interference fringes, and therefore there must be virtually no movement of the object in relation to the illuminating optics or the hologram plate between condition A and condition B (other than that deliberately induced). Equally restricting is the further requirement that there be no movement during the exposure. Additionally, movements of the order of 100 μm become difficult to deal with due to the resulting high density of fringes. (Dense fringes are difficult to discern in an image affected, unavoidably, by the speckle effect associated with coherent light.)

As a result of these restrictions, classical holographic interferometry must, with the exception of the double-pulsed technique described below, be carried out in a very stable environment, and the deliberately induced movements must be very small. This has usually entailed mounting everything on a vibration-isolated table and loading the object in some artificial way so as to produce movements that are adequately small. Techniques which compensate for whole-body motions have been developed (e.g. 'sandwich holography' [20], 'reflection holography', 'fringe control' etc). However, these cannot compensate for large deformations (as opposed to whole-body motions).

A final difficulty is the change of the very nature of the surface which may accompany some shape changes (e.g. plastic flow). It is at once the strength and weakness of holography that the random surface of a diffusely reflected object is recorded in the hologram. While it permits comparison of very small shape changes without any surface treatment, it virtually precludes comparison when the surface has changed appreciably.

The realistic loading of objects is therefore limited to very small [21] or very fragile pieces, and, as a quantitative engineering tool, the promise in principle is unlikely ever to be fulfilled in practice in other than these types of application. But then again, even there, a problem of quantification arises, since the relationships between the fringe contour interval and the displacement vector are not simple. Recent work by Stetson (e.g. [22] has improved the position on data reduction, but it is still a formidable undertaking for all but fairly simple geometries.

4.6.1.4. *Use in NDE*

Many of the objectives of NDE can be realised by the use of loads and deformations much smaller than those encountered in service, and for such purposes holography has proved to be a practical tool. The use of small loads can allow cracks and debonds and other sub-surface flaws to be detected and at least partially quantified as to size, shape and location. Also, on the presumption that the nature of the modes of vibration is not a strong function of amplitude, the vibration behaviour can be studied, and valuable information on this aspect may be obtained.

Many striking applications are described in refs [23], [24] and [25].

The small changes may be induced by direct loading, by pressurising or evacuating, by thermal expansion, or by vibrational excitation induced by piezo, magnetic or acoustic means.

Another approach which is feasible, but which has had relatively little attention, is the possibility of making a hologram of a part, then subjecting the part to service conditions for some period, and subsequently replacing it in the holographic set-up and examining for permanent change of shape induced by the in-service duty. For small parts which are quite stiff such a procedure should be possible, given a suitably rigid holding jig [21]. Luxmoore [26] has reported a similar procedure, not with the actual object and a hologram, but with two 'cast replicas' of the object's surface, one in the 'before' and one in the 'after' condition.

Optical access is an obvious requirement, and the effects of shadowing have to be considered. For inaccessible regions, Dudderar *et al.* [27] have shown that it is possible to use optical fibres; but this must be regarded as frontier, rather than established, technology.

Finally, it has, rather surprisingly, been possible to do shape comparison of two *different* objects. Reference [28] describes the production of interference fringes between two different objects; the fringe quality is, however, extremely poor and the achievement is perhaps best categorised as remarkable but of no practical significance. But if the illumination and viewing directions are grazing, then the surface behaves as a mirror; shape comparison of two different surfaces then becomes more practical (though there is a loss of sensitivity at these grazing angles). Reference [24] contains a description of the comparison of cylinder bores, and references in Vest [19] include monitoring of removal of small amounts of material.

4.6.1.5. *The various techniques*
There are two basic versions of holographic interferometry: two-wavefront superimposition and time-average (which is effectively multiple-wavefront superimposition). The two-wavefront technique can be further divided into 'double exposure' and 'real time'. 'Double exposure' itself falls into two categories, namely 'double-pulse', in which a special laser is used to fire two high energy pulses (typically nanoseconds in duration) with a very short time-lapse between them (typically microseconds), and what we might call the 'long interval' technique. The distinctions between and uses of these various methods are as follows.

In time-average, a single exposure is made lasting over many cycles of a vibrating object; the reconstruction displays the vibration pattern, and the technique is used only for this purpose.

In real-time, a first exposure is made, the hologram is developed and replaced in the original position and reconstructed so that the reconstructed image coincides with the actual object; any deformation of the object is now revealed as it occurs, as fringes over the object surface; the pattern can be photographed at any desired time. This is the most informative technique, but replacing the hologram has proved in the past to be a drawback; in-situ development overcomes this, and the introduction of dry-processing photo polymers [29] has made the technique much more convenient. Real-time can be used for both vibration analysis and general deformation.

In double-exposure, holograms of the two conditions are recorded 'blind', i.e. the fringe pattern which results is only observable after the plate has been developed; in 'long-interval', the change of shape can be controlled between the exposures, the time between each being chosen entirely for convenience; in the double-pulse technique it is extremely difficult to exercise control over the deformation, and one simply accepts what one gets as a result of, usually, an impact or an induced vibration. Double-pulse can provide information on vibration behaviour, and it is the only technique which has no requirement for rigidity between optics and specimen, the time between pulses being short enough to ensure that any movement is extremely small. 'Double-pulse' is not to be confused with a long-interval double exposure in which each exposure is made with a high-energy pulse; the name is accepted as applying only to the very-short-interval technique, using a 'double-pulse' laser.

4.6.1.6. *Considerations in relation to composites*
Factors of significance when considering holography for NDE of composite materials, in particular, are:

(i) Like all optical techniques, the only material property required is reflectance (or transmittance); the non-magnetic and inhomogeneous nature of the material is irrelevant. On the other hand, materials that are non-reflecting (e.g. many carbon-fibre composites) may present difficulties.

(ii) Composite structures tend to be less rigid than metallic structures. The corollaries of this are a requirement to induce test loads that are even further removed from 'service' loads, and a greater difficulty in re-positioning a piece after some service, or simulated service, duty. On the other hand, by the same token, sub-surface flaws are more likely to have a detectable effect at the surface.

4.6.1.7. *Practical considerations*

When contemplating the introduction of a holographic test procedure, the following should be borne in mind. A very high level of expertise is needed to produce good holograms; a different expertise is called for in assessing an interferogram. While there are virtually off-the-shelf packages [30] which can alleviate the first difficulty, they cannot help with the second. In principle, image processing is applicable to this latter end, but that particular answer is in its infancy (see Section 4.5).

There are important safety regulations to be considered [31]. These can be very restrictive if high power lasers are required, necessitating dedicated areas and equipment.

4.6.1.8. *Summary*

Holography is, in principle, a powerful tool for qualitative NDE such as flaw detection and vibration studies, and is generally applicable to composite materials. In practice, it demands the highest levels of skill, though commercial systems can help to reduce this. It is not well suited to quantitative determinations such as strain distributions. Holography is an excellent method for determining vibration patterns.

4.6.2. Heterodyne holography

For quantitative assessment of strain from holograms, it is necessary to measure the local fringe frequency. Conventionally this has been done by measuring, one way or another, the distance between two fringe centres. In heterodyne interferometry a totally different concept is used: the fringes are made to sweep across the field by introducing a slight difference in frequency between the light used for reconstructing the first exposure and that used for the second. (This is achieved by using separate reference beams for the two exposures.) A photodetector looking at a point will therefore have a sinusoidal output as fringes sweep across the point, and the output of two adjacent photodetectors will be phase-shifted one with respect to the other. A measurement of the phase-shift gives the local 'fringe spacing'.

A discrimination of 1/1000 of a fringe is entirely feasible on 'gauge lengths' much less than a fringe spacing, making heterodyning a very powerful tool for extracting detailed information from a hologram. The set-up, naturally, must differ somewhat from a conventional one; see, for example, [32].

4.6.3. Acoustical holography
4.6.3.1. *The principles*
Acoustical holography is a term referring to a means of rendering visible the interior of light-opaque objects, so that flaws can, as it were, be 'seen'. A 'point' source of a pure sinusoidal tone is the acoustic analogue of an expanding laser beam such as is used to illuminate an object for an ordinary hologram. Now, if such sound propagates through a slab of material it can be thought of as 'illuminating' the back surface which then reflects the incident sonic waves; similarly, any flaws within the thickness of the slab will also reflect the waves. If now the reflected sound meets another sonic beam (the 'reference') which is coherent with it, then a standing interference pattern will be formed, and a plane which lies within the overlap of the two will be the acoustic equivalent of the plane of a hologram.

If a record of the interference pattern in that plane were made, then, in principle, by 'replaying' it with the reference sonic beam the original sound reflected by the flaws etc. could be reproduced, and if we could actually see sound waves then we could see all the flaws, their position in three dimensions, etc. But, of course, we cannot; acoustical holography consists of the recording of the acoustic pattern and its subsequent conversion to an optical analogue, which then may be reconstructed with visible light to create a visual three-dimensional pattern corresponding to the interior of the slab on a point-by-point basis.

There are various ways in which the conversion may be carried out, including, for example, the use of a liquid surface which will develop an undulating profile in response to a standing acoustic pattern; or, again, plastics, suspensions and liquid crystals have been used. A particularly convenient technique is one wherein the plane of interest is scanned with a microphone. There is no need for a real reference wave in this case, as the electrical signal driving the sound emitter can be added electrically to the microphone signal to produce the required interference. The resultant signal is used to amplitude-modulate a narrow beam of light as it scans a photographic plate in sympathy with the microphone scan. (The scale of the scans has to be very different, due to the different wavelengths.) When developed, the plate is an optical hologram which can be reconstructed to produce the visible analogue of the back-reflected sound.

References [33] and [34] discuss many of the techniques and problems of the process.

4.6.3.2. *Relevance to composites*
There are, of course, particular considerations when insonifying a composite material, and these are dealt with in Volume 2. Within these limitations, acoustical holography may be regarded as a technique of considerable potential but not yet fully developed, despite the availability of commercial equipment [35].

4.6.4. Holographic correlation
Marom [24] reports on the use of correlation of an object surface with itself after some fatigueing process. The correlation coefficient drops with progressive fatigue damage. This technique would appear to be of considerable potential value for certain testing situations; the same positional restrictions would hold, and the process is at least as specialist as holographic interferometry.

4.7. 'SPECKLE' METHODS

4.7.1. Introduction
The phenomenon of speckle, associated with scattered coherent light, really requires to be observed to be properly appreciated. Any diffusely reflecting surface appears to be covered with myriad bright points and dark ones, lending the appearance of a roughened bright metallic surface. The effect has two distinct causes, both involving interference.

Figure 15a illustrates the first; the interference of all the scattered light arriving at points A, B, C, D results in respective brightnesses which are dependent upon the mutual phases and the intensities of all the rays going through a particular point. Since the phases are random, we can visualise that the whole space is 'filled' with little alternate light and dark volumes. This is known as 'objective' speckle; it does not depend upon the observer, and if we place a photographic plate anywhere in the space it will, upon development, be seen to be covered in tiny areas of white and black. Examination of (lensless) hologram plates reveals this type of microstructure. The larger the scattering surface the finer the speckles, since there are then higher angles between the interfering rays (on average).

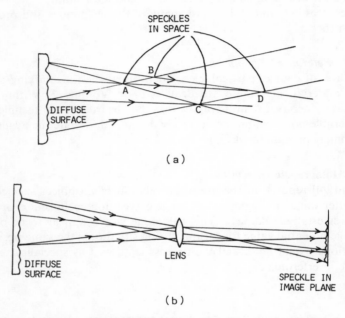

Fig. 15. (a) Formation of objective speckle; (b) formation of subjective speckle.

Figure 15b shows the second cause. Light captured by the lens is restricted by the aperture. Again, a speckle pattern results, but now the nature of the speckle depends upon the geometry of the lens; for whereas, with a large format lens, say f/2, the mean angle between rays interfering at the image plane will be high (giving fine speckles), a lens set at f/50 will produce a low mean angle and thus a coarse speckle. (If the light is not diffuse—say, for example, collimated light—then the speckle pattern is not random, and the interference pattern for a circular aperture is in the form of concentric rings, due entirely to edge-diffraction.) In diffuse light, therefore, the random pattern formed has a characteristic speckle size dependent upon the observing lens, and this is referred to as 'subjective' speckle. If the aperture is moved, the speckle pattern changes.

The important point to be made is that the nature of a speckle pattern is dependent upon two factors: the scattering properties of the observed surface (since this determines the relative phases and intensities of the rays arriving at the aperture), and the nature of the observing lens-aperture arrangement.

If now a surface changes its shape a little between two observations, so that the relative phases change, then the 'before' and 'after' speckle patterns observed by the same lens-aperture are related, and these two patterns can be used to determine the change. This is the basis of the 'speckle methods'.

Vest [19] gives a suitable treatment and refers to works of a more mathematical nature.

4.7.2. The techniques
There are two distinct techniques, each with its own set of characteristics: speckle interferometry which has a high sensitivity but low range, and speckle photography which has a higher range but a low sensitivity.

4.7.2.1. *Speckle interferometry*
In this technique, the speckle field is recorded in coherent addition with either a reference beam or another speckle field, the latter being utilised when the deformation of interest is the 'in-plane' component. Figure 16a and b show typical respective set-ups. It will be relatively readily understood that, in Fig. 16a, if the object moves towards or away from the lens (i.e. in an out-of-plane mode), then the phases of rays leaving a small area on the object change in complete harmony. Thus the object's own speckle pattern intensity remains essentially unchanged, but the phases alter. However, with the presence of the reference beam in the image plane, the *total* speckle pattern is formed by the coherent addition of the object's speckle pattern and that of the reference. The end result is that for small movements there is a contrast reversal of the speckles for every $\lambda/[2(1 + \cos \theta)] = \Delta$ say, of out-of-plane movement. With the two-beam arrangement for in-plane displacement, the mechanism is rather different in that it is not the contrast but the correlation of the speckle pattern that alters, correlation being re-established for every $\Delta = \lambda/(2 \sin \theta)$ of in-plane displacement component in the y-direction (x-displacement has no effect). By

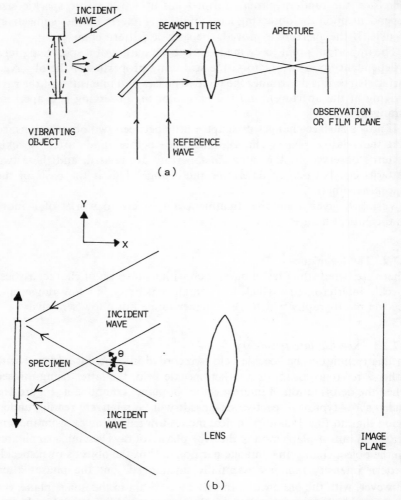

FIG. 16. (a) Measurement of out-of-plane deformation by speckle interferometry; (b) measurement of in-plane deformation by speckle interferometry.

means of double exposure, real-time or time-average methods interference fringes can be produced, showing contours of constant Δ. The sensitivity, then, is that of holographic interferometry.

Why use speckle interferometry? The major feature of speckle interferometry, in comparison with holographic interferometry, is in the recording process. Whereas holography requires the recording of minute detail, necessitating ultra-fine-grain film which, with its inherent low sensitivity, implies long exposure times, speckle interferometry on the other hand can be carried out with faster films, since the speckles can be made relatively coarse simply by reducing the aperture. It is this feature which permits the use of video-recording and electronic image comparison in ESPI (electronic speckle pattern interferometry), in which photographic processing is totally eliminated.

The problem with speckle interferometry is, strangely perhaps, the speckle! The deliberate coarsening of the speckle, taken together with different mechanisms of fringe formation, results in fringes that are, by comparison with conventional holographic fringes, very noisy, and thus the tolerable density of fringes is more limited. Parks [36] has examined many published speckle interferograms and has concluded that the range is very restricted.

Speckle interferometry may therefore be classified as more convenient than holographic interferometry, but suited only to the measurement of very small movements (typically several micrometres). Lokberg [37] has applied the principle of heterodyning to achieve a resolution of a few nanometres.

4.7.2.2. *Speckle photography*
Speckle photography consists of simply photographing a coherently illuminated object without a reference beam. If this is done before and after deformation as a double-exposure, then the two recorded sets of speckle patterns will, depending upon the amount of movement, be well correlated. Any movement along the optical axis changes the object's speckle pattern intensity but little, as explained above. Now, since there is no reference beam to introduce phase-related changes, the two recorded speckle patterns will be virtually identical. However, in-plane movement causes the speckles to shift 'sideways', by an amount proportional to the magnification of the image and to the movement. We therefore have two nominally identical speckle patterns but with corresponding speckles actually shifted with respect to each other according to the local in-plane displacement on the object.

For typical strain gradients, it will be understood that in a restricted locality (e.g. 1 mm diameter) all speckles will have moved by virtually the same amount. If now we illuminate such an area on the developed plate with a raw laser beam, each pair of speckles will act as two point sources, and Young's fringes will be formed by diffraction (see Fig. 12). This is the basis of the speckle photography system; reference is made to Vest [19] for further detail. A speckle photograph is usually referred to as a specklegram.

The pitch of the Young's fringes gives the magnitude of the local displacement vector, and their inclination describes its direction. By scanning the plate, the whole in-plane displacement field can be mapped, and quantities such as strain may then be extracted. This is a time-consuming process which can, in principle, be automated to any desired degree (see for example refs [17] and [38]).

Whole-field pictures having contours of constant displacement component in any desired direction can be obtained directly by illuminating the whole specklegram by a collimated beam and observing it through an aperture suitably placed in the focal plane of a lens. However, the fringes are, again, of very poor contrast.

If during the exposure the plate is positioned in the focal plane of the lens, rather than in the image plane, then the speckles will shift due to whole-body tilt. The corollary of this fact is that if the image is poorly focused, then the speckle movement will be a mixture of in-plane deformation and tilts, and this renders somewhat problematical the use of the technique for non-flat objects.

For flattish objects, it is probably true to say that speckle photography is the most practical of the non-grid coherent systems, at least for displacement and strain measurement, due to its tolerance of considerable object movements (e.g. 1 mm). The upper limit of strain is about 2000 microstrain due to the concomitant change in the speckle pattern as the nature of the surface itself is altered significantly. The lower limit of range is determined by the necessity that the speckles move more than their own 'diameter', i.e. they do not overlap. In principle this can be as low as 5 μm, though the depth-of-focus difficulty may increase this substantially (e.g. to 100 μm) since sharp focus involves small aperture and thus large speckle. However, for strain measurements this is unimportant since it is *differential* movement which is of interest, and sufficient deliberate whole-body movement is readily arranged so as to ensure that the speckles do not overlap. The sensitivity of the method is dependent upon the discrimination with which the pitch of the Young's fringes, which tend to be fairly speckly, can be measured. Stetson [39] has used heterodyning with two single-exposure photographs to increase the sensitivity by two orders of magnitude compared to centre-of-fringe estimation methods. Speckle photography cannot be used in real-time or time-average modes.

4.7.3. Relevance of speckle methods to NDE of composites
Composites have no particular significance in relation to speckle methods any more than to holography. In certain situations, e.g. when the induced movements are particularly small, speckle interferometry, especially in the electronic form, can be a convenient alternative to holography. Where quantitative point-by-point information on in-plane movement could be useful, then speckle photography may be of interest.

4.7.4. Practical considerations
Just as with holography, a considerable specialist expertise is, in general, required for both the use of speckle systems and the subsequent interpretation. Proprietary self-contained ESPI systems are available commercially [40].

The facilities required, the safety considerations, etc., are also similar to those involved in holography.

4.8. SHEAROGRAPHY

4.8.1. Description
The term Shearography was introduced by Hung [41] to describe a particular form of 'speckle-shearing interferometry' (SSI), other embodiments of which are described in Vest [19]. SSI consists of causing two coherent images of the object surface to be formed simultaneously, but with one *shifted* with respect to the other, and then making a recording. After some deformation has taken place, a second recording is made in the same way, in the double-exposure manner. By fairly straightforward optical processing, fringes are observed, these fringes being contours of the derivative of displacement rather than of displacement itself (e.g., if the displacement sensitivity is to the out-of-plane, then the fringes are contours of constant slope).

4.8.2. Advantages
The advantage of SSI is a decreased mechanical stability requirement; in fact the technique is insensitive to whole-body motions and reacts only to differential movements within the body. This insensitivity originates in the fact that the speckle-forming rays for any locality in the image are arriving from two different localities of the object. This leads to a double-structure in the speckles, such that each 'speckle' is composed of a finer substructure of speckles which approximates to fine fringes. Whole-body

movements do not alter the macrostructure appreciably, and the substructures do not change their fringe spacing but only their fringe positions—and this in sympathy for all speckles in the image. Differential movement, however, causes the fine fringes to change their positions differently within the different speckles, and by suitable post-processing the derivative-of-displacement contours can be produced.

4.8.3. Limitations
Although the method is insensitive to whole-body motion, there are restrictions on the amount that can be tolerated before the method becomes ineffective. In particular, the in-plane movement, whether whole-body or differential, must be such that the macro-speckles do not move more than about half of one speckle 'diameter'. Within these limitations, however, whole-body movement is irrelevant, provided a fast enough exposure is used to ensure that there is no 'smearing' of the recorded speckle pattern.

4.8.4. Practical considerations
Generally these will tend to be similar to those for holography and speckle methods, but with considerable easing of the rigidity constraints. Interpretation is considerably facilitated by the lack of any fringes due to whole-body motion. Impressive exemplification on car tyres is given in [41].

It is not possible to have real-time shearography; it is intrinsically a process requiring post-treatment to reveal the fringes, though, at least in principle, it is conceivable that this could be done using video with electronic processing. Reference [41] indicates that commercial equipment may be available.

4.9. BRITTLE LACQUER

4.9.1. The principle
The technique of applying a thin brittle coating to the specimen and then applying a service load has been widely accepted for many years as a valuable aid in the identification of areas of high strain. Indeed, it is routinely used in the proof-testing of pressure-vessels [42].

It is in essence a very simple technique. The surface is coated with a uniform thin film of the lacquer, which is solvent-based. The application may be by brush or spray. Upon drying, the coating develops its brittle characteristic, so that when the specimen is loaded, the lacquer will crack when the surface strain exceeds a certain level known as the threshold level. (Various lacquers are available, having different thresholds.) Thus, by proper choice of threshold, the surface will be seen, after loading, to have cracks in the areas of high strain, indicating both their localities and the strain magnitude.

4.9.2. The practicalities
The method requires considerable care and attention in the application and curing of the coating. In particular, the uniformity of thickness and the temperature and humidity during curing are most important. Consequently the technique is regarded as semi-quantitative, but despite this, its simplicity has made it popular. Reference [43] gives a suitable introduction to the method and contains further references for more detail.

4.9.3. Relevance to NDT of composites
Since many composites are based on polymeric matrices, the compatibility of the solvent with the matrix will require special consideration. Given this proviso, there is no reason why the method should not be applicable. Indeed, the authors have observed such a test on glass-reinforced-polyester pipework joints, with informative results.

4.10. PHOTOELASTIC COATING

4.10.1. The principle
The photoelastic effect is widely known and used for the testing of models of service equipment, the model being fabricated of epoxy or some other material exhibiting the birefringent effect. This is the phenomenon whereby the velocity of light through the material is altered according to the stress in the material and the polarisation of the light. The output is in the form of fringes which indicate the stress levels and directions.

For the testing of the actual components themselves, a coating technique is used. For complex shapes, the coating is applied by heat-softening a piece of the birefringent material and moulding it carefully to the form of the component. Reference [44] gives appropriate and practical guidance.

4.10.2. Practical considerations
A problem with the coatings is that they may require to be rather thick to provide the sensitivity required (usually of the order of 1.5 mm, sometimes

up to 6 mm). There is also a difficulty in interpreting the results when the surface is being observed from a direction not close to the perpendicular.

Finally, it should be pointed out that the stress information is obtained as the difference of the principal stresses, and that the methods of separation of the stresses can be involved and inaccurate, except close to free edges.

4.10.3. Relevance to NDT of composites

Whereas a thickish coating on a metallic specimen may have an insignificant reinforcing effect, the same coating on a specimen of composite material might itself alter the prevailing stresses. With that caveat, together with a consideration of the compatibility of the adhesive employed, the method is as applicable to composites as to homogeneous materials. Reference [45] contains exemplification on graphite-epoxy laminae, the coating being 0.25 mm thick.

4.11. OPTICAL FIBRE CRACK MONITOR

4.11.1. The principle

If an optical fibre is bonded to a specimen, then a crack in the specimen will, as it opens, tend to break the fibre. If light is being shone down the fibre, then the break will be detectable, either by simple observation which will reveal a pinpoint of light emanating from the break, or by monitoring the light level emitted at the end of the fibre—a large attenuation occurring due to the break. This then is the basis of the method described by Boyle [46].

4.11.2. The practicalities

The method probably relates more to testing in service, or in a service simulation, rather than to a routine quality-of-manufacture requirement. However, it could be of interest in ascertaining the criticality of an already detected crack; for, depending upon whether a crack actually opens or not when the component is under load, it may be possible to make a decision as to the serviceability of the component.

As reported in ref. [46], crack openings of 20 μm are detectable, and it is hoped to bring this down to 10 μm. The technique is very simple, entirely practical, and there is comment in [46] as to commercially available equipment; see also ref. [47]. For monitoring crack growth, a number of fibres are laid side by side, and the progression of the crack is indicated by sequential breaking of the fibres above it.

REFERENCES

1. P. W. Ford, Optical probe, *1972 Annual Meeting of the Optical Society of America,* Paper WE12; abstracted in JOSA, 1972, **62**, 1341.
2. G. S. Ludwig, F. C. Brenner and C. Conley, Automatic tyre tread gauging machine, Tech. Dept., US Dept of Transportation, Nat. Highway Safety Admin., Washington, DC, Nov. 1977, Report DOT-HS-802831.
3. P. Petit, G. Tourscher, M. Machet and M. Kassel, Shape Measurement for hot rolled strip, *SPIE*, Vol. 164, *Proc. 4th European Electro-optics Conf.*, Utrecht, 1978, p. 160.
4. Diffracto Ltd, 6360 Hawthorne Drive, Windsor, Ontario, Canada.
5. A. E. Ennos and M. S. Virdee, Precision measurement of surface form by laser autocollimation, *Proc. SPIE*, Vol. 398, Industrial Applications of Laser Technology, Geneva, 1983, p. 252.
6. D. P. Himmel, A laser-based system for automatic industrial inspection, *Proc. 4th Int. Joint Conf. on Pattern Recognition*, Kyoto, Japan, 1978; IEEE, New York, 1979, pp. 952–954.
7. H. Takasaki, Moiré topography, *Applied Optics*, June 1970, **9**(6), 1467–1472.
8. Y. Yoshino, M. Tsukiji and H. Takasaki, Moiré topography by means of a grating hologram, *Applied Optics*, October 1976, **15**(10), 2414–2417.
9. G. T. Reid, R. C. Dixon and H. I. Nesser, Absolute and comparative measurements of three-dimensional shape by phase measuring moiré topography, *Optics & Laser Technology*, December 1984, 315–319.
10. A. Nordbryn, Moiré topography with a charge coupled device TV camera, *Proc. SPIE Int. Soc. Opt. Eng. (USA)*, (1983), **398**, 208–13: Industrial Applications of Laser Technology, Geneva.
11. D. Post and T. F. MacLaughlin, Strain analysis by moiré fringe multiplication, *Experimental Mechanics*, September 1971, **11**, 408.
12. J. Cargill, Measuring strains under welds by the moiré fringe technique, *Strain*, January 1970, **6**(1), 28.
13. J. M. Burch and C. Forno, High resolution moiré photography, *Optical Engineering*, 1975, **14**, 178–185.
14. C. A. Walker, J. McKelvie and A. McDonach, Experimental study of inelastic strain patterns in a model of a tube–plate ligament using an interferometric moiré technique, *Experimental Mechanics*, 1983, **23**, 21–29.
15. D. Post and W. A. Baracat, High-sensitivity moiré interferometry—a simplified approach, *Experimental Mechanics*, 1981, **21**, 105–110.
16. A. McDonach, J. McKelvie, P. MacKenzie and C. A. Walker, Improved moiré interferometry and applications in fracture mechanics, residual stress and damaged composites, *Experimental Techniques*, June 1983, **6**(6), 20–24.
17. D. W. Robinson, Automatic fringe analysis with a computer image-processing system, *Applied Optics*, 15 July 1983, **22**(14), 2169–2176.
18. T. G. F. Gray, J. McKelvie, P. MacKenzie and C. A. Walker, Interferometric measurement of *J* for arbitrary geometry and loading, *Int. J. Fracture*, 1984, **24**, 109–114.

19. C. M. Vest, *Holographic Interferometry*, John Wiley & Sons, 1979.
20. N. Abramson, Sandwich holography: an analogue method for the evaluation of holographic fringes, p. 631 in ref. [19].
21. See, for example, Discussion, pp. 77–78 in ref. [19].
22. K. A. Stetson, Use of projection matrices in holographic interferometry, *J. Opt. Soc. America*, 1979, **69**, 1705–1710.
23. E. R. Robertson (ed.), *Proc. Conf., The Engineering Uses of Coherent Optics*, Cambridge University Press, 1976.
24. R. K. Erf (ed.), *Holographic Nondestructive Testing*, Academic Press, 1974.
25. W. F. Fagan (ed.), *Proc. Conf. Industrial Applications of Laser Technology*, SPIE, Vol. 398, Geneva, 1983.
26. F. A. A. Amin and A. Luxmoore, A transmission replica system for strain measurement, *NDT International*, June 1979, **12**(3), 115–120.
27. T. D. Dudderar, J. A. Gilbert, R. A. Franzel and J. H. Schamell, Remote vibration measurement by time averaged holographic interferometry, *Proc. V Int. Congress on Experimental Mechanics*, Montreal, June 1984, p. 362.
28. Z. Fuzessy and F. Gyimesi, Difference holographic interferometry, p. 240 in ref. [24].
29. J. Schorner and H. Rottenkolber, Industrial application of instant holography, p 116 in ref. [24].
30. Information available from Rottenkolber Holo-System GmbH, Erhardstrasse 2, 8000 Munchen 5, West Germany, and Newport Corporation, 18325 Mt Baldy Circle, PO Box 8020, Fountain Valley, CA 92728-8020, USA.
31. BS 4803, *Radiation safety of laser products and systems*, Part 3, *Guidance for users*, British Standards Institution, 1983.
32. R. Dandliker, B. Eliasson, B. Ineichen and F. M. Mottier, Quantitative determination of bending and torsion through holographic interferometry, p. 99 in ref. [3].
33. R. W. B. Stephens and H. G. Leventhall (eds), *Acoustics and Vibration Progress*, Vol. 2, Chapman & Hall, 1976.
34. B. B. Brenden and H. D. Collins, Acoustical holography with scanned hologram systems, Chap. 10 in ref. [22].
35. I. G. Scott and C. M. Scala, A review of non-destructive testing of composite materials, *NDT International*, April 1982, **15**(2), 75–86.
36. V. J. Parks, The range of speckle metrology, *Experimental Mechanics*, 1980, **20**(6), 181–191.
37. O. J. Lokberg and G. A. Slettemoen, Some industrial applications of electronic speckle pattern interferometry, p. 295 in ref. [24].
38. G. E. Maddux, R. R. Corwin and S. L. Moorman, An improved automated data-reduction device for speckle metrology, *Proc. 1981 Spring Meeting*, SESA, Dearborn, Michigan.
39. K. A. Stetson, The use of heterodyne speckle photogrammetry to measure high temperature strain distributions, pp. 46–55 in *SPIE*, Vol. 370, *Holographic Data in Non-destructive Testing*, Dubrovnic, 1982.
40. Information available from Ealing Electro-Optics, Watford, UK, and Newport Corporation, Fountain Valley, California, USA.
41. Y. Y. Hung and R. M. Grant, Shearography: a new optical method for non-destructive evaluation of tires, *Rubber Chemistry and Technology*, 1981, **54**(5), 1042–1050.

42. BS 5500, *Fusion welded unfired pressure vessels*, Section 5.8, *Hydraulic testing*, British Standards Institution.
43. I. B. MacDuff (ed.), *Methods and practice for stress and strain measurement*, Part 3, *Optical methods for determining strain and displacement*, British Society for Strain Measurement, Monograph, 1978.
44. P. Stanley (ed.), *Methods and practice for stress and strain measurement*, Part 2, *Photoelasticity*, British Society for Strain Measurement, Monograph, 1977.
45. I. M. Daniel, Mixed mode failure of composite laminates with cracks, *Proc. V Int. Congress on Experimental Mechanics*, Montreal, 1984, pp. 320–327.
46. H. B. Boyle, 'Real time' crack detection by the use of optical fibres bonded to the surface of materials, *Proc. V Int. Congress on Experimental Mechanics*, Montreal, 1984, pp. 164–167.
47. Information available from National Maritime Institute Ltd, Feltham, Middlesex, UK.

Chapter 5

Vibration Techniques

P. CAWLEY

Department of Mechanical Engineering, Imperial College, London, UK

and

R. D. ADAMS

Department of Mechanical Engineering, University of Bristol, UK

5.1. INTRODUCTION

Vibration techniques for non-destructive testing have been used for hundreds, probably thousands, of years, yet the subject is still in its infancy. A standard technique for testing earthenware cooking pots has always been to tap them and listen to the ring. A good pot will produce a sustained, clear note while a cracked pot will sound 'dead'. The same technique is used in the crystal glass industry. The railway wheel tapper used to walk along the train tapping each wheel in turn. Again, cracked wheels did not ring for as long as good ones. The test, which in this paper is termed the 'wheel-tap' test, is 'global' since the whole component is tested by a tap at a single point. It is therefore a very rapid inspection method.

A superficially very similar technique is regularly used for testing laminated structures such as bonded panels. This is the 'coin-tap' test which involves tapping each part of the panel with a coin. Again, a defective area sounds 'dead'. Defects such as adhesive disbonds, delaminations in composite materials and defective honeycomb construction can be detected by a skilled operator. The method is not suitable for the detection of transverse cracks (i.e., cracks running normal to the surface which is tapped). The coin-tap test is frequently confused with the wheel-tap technique but operates on a quite different principle. The coin-tap test is a 'local' test which will only detect defects at the location of the tap. Therefore if the whole structure is to be tested, each part of it must be tapped and the test is much more time-consuming than the wheel-tap method. However, the local test is more sensitive to certain types of defect and the two tests have quite different areas of application.

In the wheel-tap test, the operator detects differences in the pitch (frequency) and decay rate (damping) of the sound produced when good and defective wheels are struck. The test therefore depends on changes in the natural (resonant) frequencies and damping of the wheel with damage. These are properties of the whole wheel and do not depend on the position of excitation. Analysis of the vibration or sound waveform produced by a tap is a very attractive means of obtaining these properties but, unfortunately, the human ear cannot achieve great accuracy or reliability and, until recently, the electronic equipment required to carry out the test was prohibitively expensive and bulky. Apart from a few specialist cases, the test has therefore remained subjective which has severely limited its application.

There are many other methods for measuring natural frequencies and damping, some of which have been applied to non-destructive testing. However, in most cases, they are not as quick and convenient as the tap test and they have not found widespread use. With the advent of micro-electronics, the cost and size of the equipment required to extract the natural frequencies and damping from the sound produced by a tap has been greatly reduced. It is anticipated that this will lead to an upsurge of interest in the application of the technique to both composite and metal structures.

After reviewing the vibration properties of composite materials and structures, this paper discusses different methods for measuring vibration properties. Reports of 'global' tests based on natural frequency and damping measurements are discussed in Section 5.5. These are not classified according to the particular method used for measuring the properties.

The advent of micro-electronics has also produced new opportunities in the field of local measurements, which are discussed in Section 5.6. For example, an automated version of the coin-tap test has been developed. A variety of local properties may be measured, some of which require excitation at a single point of the structure, measurements being taken at each point of interest, while others require both excitation and measurement at each test point. Section 5.6 is divided into these categories.

The tests described here use frequencies of vibration which are mainly in the sonic range, which extends up to 20 kHz. Higher frequency ultrasonic methods will be discussed in Volume 2.

5.2. VIBRATION PROPERTIES OF COMPOSITE MATERIALS

5.2.1. Introduction

The vibration properties which concern us are the damping and the dynamic modulus. These are defined in Fig. 1. When taken round a stress cycle, all materials show a non-singular relationship between stress and strain. The modulus is given by the mean slope of the stress–strain loop. For most materials there is little ambiguity in this definition, since the loop is almost indistinguishable from a straight line. The area, ΔW, of the loop represents the work done against 'internal friction' and is the amount of energy dissipated during the cycle.

Referring to Fig. 1, it can be seen that the maximum strain energy

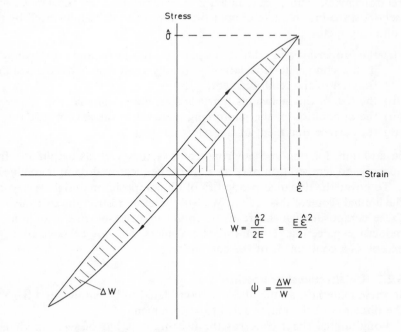

FIG. 1. Definition of specific damping capacity ψ.

stored per unit volume in the cycle is $W = \hat{\sigma}^2/2E = E\hat{\varepsilon}^2/2$. We now define the specific damping capacity, ψ, of the material as

$$\psi = \frac{\Delta W}{W} \tag{1}$$

This quantity is usually expressed as a percentage.

ψ is related to other commonly used damping parameters by the relationships

$$\psi = \frac{2\pi}{Q} = 2\delta = 2\pi\eta = 4\pi c = 2\pi\left(\frac{\omega_2 - \omega_1}{\omega_n}\right) \tag{2}$$

where Q = quality (amplification) factor, δ = logarithmic decrement, η = loss factor, c = proportion of critical damping, ω_n = the natural frequency, and ω_1, ω_2 = the half power points (see Section 5.4).

The main sources of internal damping in a composite material arise from microplastic or viscoelastic phenomena associated with the matrix and relative slipping at the interface between the matrix and the reinforcement. Thus, excluding the contribution from any cracks and debonds, the internal damping of the composite will be influenced by the following factors:

(i) the properties and relative proportions of matrix and reinforcement in the composite (the latter is usually represented by the volume fraction of the reinforcement, v_f);
(ii) the size of the inclusions (particle size, fibre diameter, etc.);
(iii) the orientation of the reinforcing material to the axis of loading;
(iv) the surface treatment of the reinforcement.

In addition, loading and environmental factors such as amplitude, frequency and temperature may also affect the measured damping values.

To cover the dynamic properties of all composite materials is beyond the limited scope of this review. We will therefore restrict the treatment to those composites which are to be found in stress-bearing situations in modern engineering. Unreinforced polymers will not be covered here, except as a component of the composite.

5.2.2. Unidirectional composites
In these materials, all the fibres are considered to be parallel and lying in the direction of the major axis of the specimen.

Longitudinal shear concerns the twisting of a bar of such an aligned composite. Thus the longitudinal shear modulus of CFRP and GFRP is principally a function of the matrix shear modulus, the fibre shear modulus and the volume fraction of fibres. None of the existing micromechanics theories accurately fit the experimental data [1], but the numerical prediction of Adams and Doner [2] gives good agreement. From such a curve as is shown in Fig. 2 it is possible to determine the volume fraction v_f if the matrix shear modulus G_m is known, and vice versa.

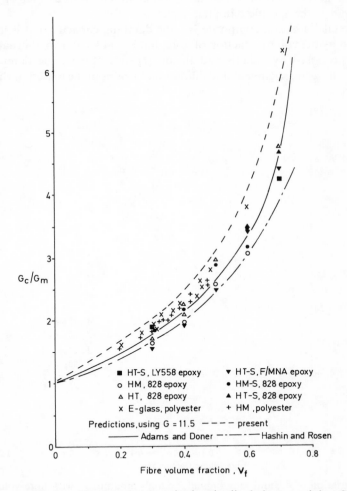

FIG. 2. Variation of reduced composite longitudinal shear modulus G_c/G_m with fibre volume fraction v_f. (N.B: $G_{LT} = G_c$.)

For longitudinal shear loading, Hashin [3] has shown that for viscoelastic materials

$$\psi_{LT} = \frac{\psi_m(1-v_f)[(G+1)^2 + v_f(G-1)^2]}{[G(1+v_f)+1-v_f][G(1-v_f)+1+v_f]} \tag{3}$$

where the suffices f and m mean fibre and matrix, v_f is the volume fraction of fibre, G is the ratio of the shear modulus of the fibre to that of the matrix, and ψ the specific damping capacity.

This result leads to a composite specific damping capacity that is little influenced by the volume fraction of fibre, but both this and an alternative solution proposed by Adams and Bacon [1] overestimate the damping because of fibre misalignment and dilatational strains in the material which

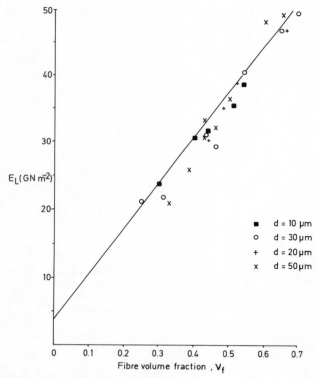

FIG. 3. Variation of flexural longitudinal Young's modulus E_f with fibre volume fraction v_f for glass fibres of different diameters d. □, $d = 10$ μm; +, $d = 20$ μm; ○, $d = 30$ μm; ×, $d = 50$ μm; ——, law of mixtures.

contribute little to the damping but significantly to the stored strain energy.

The longitudinal Young's modulus, E_L (the tensile modulus in the direction of the fibres in a unidirectional composite), is given by the rule of mixtures and is

$$E_L = E_f v_f + E_m(1 - v_f) \tag{4}$$

where E_f, E_m are the fibre and matrix moduli and v_f is the volume fraction of fibre. This rule is well-obeyed experimentally (see Fig. 3 and refs [1],[4]), and may be used as a check on any parameter provided the others are known. This relationship was derived for normal axial loading (tension or compression), but also applies in flexure provided shear effects can be neglected. It is also possible to predict the damping capacity of unidirectional material when stressed in the fibre direction by using the law of mixtures and assuming that all the energy dissipation occurs in the matrix. On this basis, we arrive at the equation

$$\psi_L = \psi_m(1 - v_f) E_m/E_L \tag{5}$$

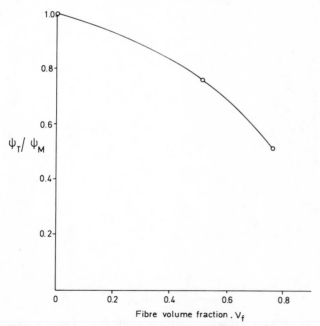

FIG. 4. Variation of the ratio of transverse damping ψ_T to matrix damping ψ_m with fibre volume fraction v_f.

where E is Young's modulus and the suffix L means longitudinal tensile/compressive properties of the composite.

However, it is found that this expression underestimates considerably the measured value of ψ_L, even when considerable effort has been made to eliminate extraneous losses. There are several contributions to the discrepancy. First, the smaller the fibre diameter, the larger is the surface area of fibre per unit volume. Adams and Short [4] showed that, for glass fibres of 10, 20, 30 and 50 μm diameter in polyester resin, there was a consistent increase in ψ_L with reduction in fibre diameter. Second, the problem of misalignment is not insignificant, as is shown below for angle-ply composites. Third, any structural imperfections, such as cracks and debonds, lead to interfacial rubbing and hence to additional losses. Finally, although the effect of shear is usually negligible in stiffness measurements, this is less true for the damping.

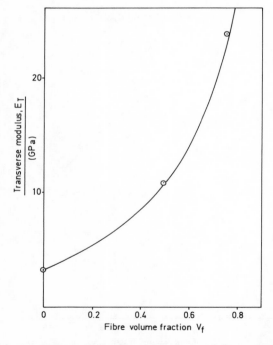

FIG. 5. Variation of the transverse modulus E_T with fibre volume fraction v_f of GFRP in flexure. ———, Halpin and Tsai's theoretical curve; $E_{ft} = 70$ GPa, $E_m = 3.21$ GPa; ⊙ = experimental points.

The damping in longitudinal shear, ψ_{LT}, is of the order of 50–100 times larger than the tension/compression component. Thus, although only a few per cent of the energy is stored in shear, this can make a substantial contribution to the total predicted value.

In the transverse direction, the damping is, as in shear, very heavily matrix dependent. Also, there is again no reliable micromechanics theory for predicting ψ_T. Experiments covering a wide variety of fibres (from E-glass to HM carbon) showed that the transverse damping is largely independent of both fibre type and surface treatment. Volume fraction does, in a similar way to longitudinal shear, have a significant effect on ψ_T. Some experimental results to illustrate this point are given in Fig. 4. Similarly, Fig. 5 shows that the transverse Young's modulus E_T increases markedly with volume fraction.

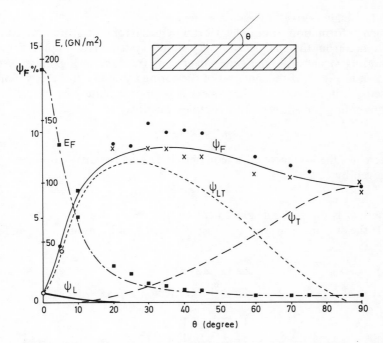

FIG. 6. Variation of flexural Young's modulus E_F and damping ψ_F with fibre orientation θ for HM-S carbon fibre in DX209 epoxy resin for a fibre volume fraction of 0.5. ●, ×, ψ_F for two similar but separate plates; ○, ψ_F in vacuo; ■, average E_F for plates 1 and 2 (values virtually coincident). Theoretical predictions: ––––, E_F; ——, ψ_F total.

Figure 6 shows the theoretical and experimental values of the overall damping, ψ_F, of an off-axis carbon fibre beam, together with the separate theoretical contributions from stresses in the L, T and LT directions. Figure 6 includes the full theoretical expression for the variation of damping with angle θ and also shows the separate contributions from direct stresses in the direction of the fibres (ψ_L), transverse to the fibres (ψ_T) and in shear (ψ_{LT}). The theoretical prediction and experimental measurement of the variation of Young's modulus, E_F, with angle is also shown in Fig. 6. Excellent agreement between the theory and experiment is shown for modulus and damping.

5.3. VIBRATION CHARACTERISTICS OF STRUCTURES

5.3.1. Single degree of freedom system

Much information about the vibration behaviour of structures can be obtained from the analysis of a very simple system. The simplest form of vibrating structure is the single degree of freedom spring–mass system shown in Fig. 7. If the material of the spring exhibits a hysteresis loop as discussed in Section 5.2, the system may approximately be modelled by representing the stiffness as a complex quantity.

$$k^* = k(1 + i\eta) \tag{6}$$

where η is the loss factor of the material. The energy dissipation per cycle is given by

$$\Delta W = \pi \eta k u_0^2 \tag{7}$$

where u_0 is the cyclic displacement amplitude.

If the system shown in Fig. 7 is subjected to a harmonic exciting force

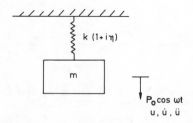

FIG. 7. Single degree of freedom system with hysteretic damping in forced vibration.

$P_0 \cos \omega t$, the displacement response will be given by $u = u_0 \cos(\omega t - \varphi)$ where

$$\left|\frac{u_0}{P_0}\right| = \frac{1}{k((1-r^2)^2 + \eta^2)^{\frac{1}{2}}} \tag{8}$$

where $r = \omega/\omega_n$ = frequency ratio, ω = frequency in rad/s ($\omega = 2\pi f$), $\omega_n = \sqrt{(k/m)}$ = natural frequency, and m = mass. The phase angle, φ, is given by

$$\tan \varphi = \frac{\eta}{1-r^2} \tag{9}$$

The maximum response occurs at the natural frequency ($r = 1$) and is given by

$$\left|\frac{u_0}{P_0}\right|_{max} = \frac{1}{k\eta} \tag{10}$$

A typical frequency response function, $|u_0/P_0|$ versus frequency, is shown in Fig. 8. If the damping is increased, the resonant frequency is not changed but the maximum response is reduced. If the phase angle between force and response is also measured, the results may be presented in a Nyquist plot. At any frequency, ω, the response, u_0/P_0, is plotted as a

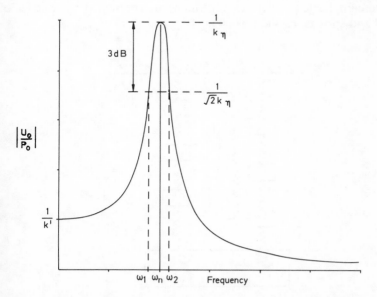

FIG. 8. Frequency response function for system of Fig. 7.

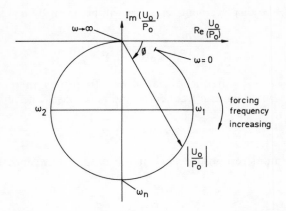

Fig. 9. Nyquist plot of frequency response function of system shown in Fig. 7. ω_n is natural frequency; ω_1 and ω_2 are half-power points.

vector of length $|u_0/P_0|$ at an angle θ to the reference axis. This results in a circular locus, as shown in Fig. 9, whose diameter is inversely proportional to the damping factor, η.

More complicated structures can often be represented by a series of masses and springs, as shown in Fig. 10.

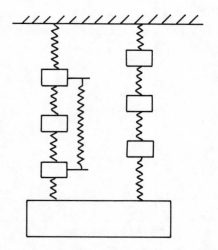

Fig. 10. Multi-degree of freedom system.

Frequency response information is often presented in terms of the impedance which is defined as

$$Z = \frac{P_0}{v_0} \tag{11}$$

where v_0 is the velocity amplitude corresponding to a harmonic exciting force of amplitude P_0. For harmonic motion,

$$v_0 = \omega u_0 \tag{12}$$

so

$$Z = \frac{P_0}{\omega u_0} \tag{13}$$

Hence, resonances correspond to minima of an impedance versus frequency plot.

5.3.2. Continuous structures
5.3.2.1. *Types of vibration*

The system discussed above has only one degree of freedom and so has only one mode of vibration. Therefore the frequency response curve shows a single resonant peak. Continuous structures such as beams, plates and shells can vibrate in different ways, for example axial, torsional and flexural vibration of a beam or shaft, and each type of vibration theoretically has an infinite number of modes. Therefore the frequency response curve of a continuous structure has many resonant peaks, as shown in Fig. 11. However, around each peak, the motion is dominated by a single mode and so the response may be regarded as the sum of a series of single degree of freedom responses. This concept can be used in the measurement of the properties of each mode, as discussed in Section 5.4.

The simplest case is the uniform bar in axial vibration, for which the axial displacement, u, of a section distance x from one end is governed by the equation

$$c^2 \frac{\partial^2 u}{\partial x^2} = \frac{\partial^2 u}{\partial t^2} \tag{14}$$

where c is the velocity of extensional waves (sound waves) and equals $\sqrt{(E/\rho)}$, where E is the effective Young's modulus in the direction of wave propagation (this will vary with fibre orientation in a composite beam) and ρ is the density. The steady-state solution to this equation is of the form

$$u = (A \cos \alpha x + B \sin \alpha x)(C \cos \omega t + D \sin \omega t) \tag{15}$$

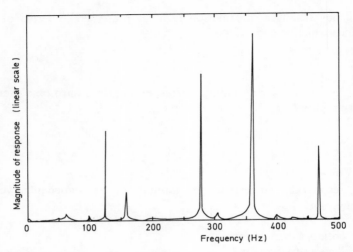

Fig. 11. Frequency response of rectangular aluminium plate (from Cawley and Adams [5]).

where A and B are constants determined by the spatial boundary conditions, C and D are constants determined by the time boundary conditions, and $\alpha = \omega/c$. For a bar with free ends resonating in its fundamental (lowest) mode of vibration, we have

$$u = u_0 \cos\left(\frac{\pi x}{l}\right) \qquad (16)$$

where u_0 is the cyclic displacement at $x = 0$ and l is the length of the bar. This defines the displacement *mode shape* and is shown in Fig. 12a. Note that the displacement is plotted vertically for visual convenience even though the motion is parallel to the axis of the bar. Nodes occur where the displacement is zero.

Sometimes, we are concerned with the cyclic stresses or strains. The strain, ε, is defined by $\varepsilon = \partial u/\partial x$ and the stress, σ, is defined by $\sigma = E\varepsilon = E\partial u/\partial x$ (assuming zero damping, although the error in most practical cases is negligible). In Fig. 12b we can see that the mode shapes for strain are quite different from those for displacement. For the first free–free mode, differentiation gives

$$\varepsilon = \frac{\partial u}{\partial x} = -\frac{\pi u_0}{l} \sin\left(\frac{\pi x}{l}\right) \qquad (17)$$

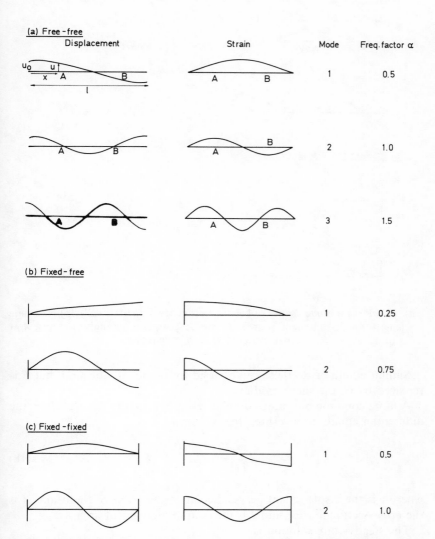

FIG. 12. Displacement and strain (stress) mode shapes for axial vibration of uniform bars. Frequency = $\alpha c/l$ Hz, where $c = \sqrt{(E/\rho)}$ and l = length; E = Young's modulus, ρ = density.

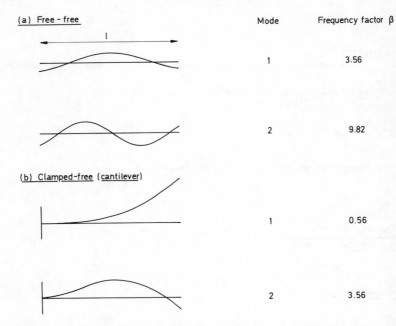

FIG. 13. Flexural mode shapes of beams. Frequency = $(\beta/l^2)\sqrt{(EI/A\rho)}$ Hz where l = length, I = 2nd moment of area of cross-section, E = Young's modulus, A = cross-sectional area, ρ = density.

Similar equations to these can be used for torsional vibration, but E is replaced by G, the shear modulus.

A more common but more complicated case is that of flexure. Here, the differential equation governing the motion is

$$\rho A \frac{\partial^2 v}{\partial t^2} = -EI \frac{\partial^4 v}{\partial x^4} \qquad (18)$$

where v is the displacement perpendicular to the plane of the beam, A is the cross-sectional area, and I is the second moment of area.

The steady-state solution is

$$v = (P\cos\alpha x + Q\sin\alpha x + R\cosh\alpha x + S\sinh\alpha x)(T\cos\omega t + U\sin\omega t) \qquad (19)$$

where P, Q, R and S are constants governed by the spatial boundary conditions, T and U are constants governed by the time boundary conditions, and

$$\alpha^4 = \frac{\rho A \omega^2}{EI} \tag{20}$$

Typical mode shapes and natural frequencies for free–free and fixed–free boundary conditions are shown in Fig. 13. The fixed–free condition is similar to the situation experienced by a tuning fork or a turbine blade.

5.3.2.2. *Beams cut from laminated plates*
Beams and plates are often made from a succession of laminae. For these, it is necessary to use the theory of laminated plates and to evaluate the contributions to damping made by each layer. Beams are a special case of plates, but are best treated separately because the theory of vibrating beams is much easier than that of plates. In any case, the theory can only be outlined in a review of this length.

$$\begin{bmatrix} \sigma_1 \\ \sigma_2 \\ \sigma_6 \end{bmatrix}_k = \begin{bmatrix} Q_{11} \, Q_{12} \, Q_{16} \\ Q_{12} \, Q_{22} \, Q_{26} \\ Q_{16} \, Q_{26} \, Q_{66} \end{bmatrix}_k \begin{bmatrix} \varepsilon_1 \\ \varepsilon_2 \\ \varepsilon_6 \end{bmatrix}_k \tag{21}$$

where the values Q_{ij}^k are the stiffness matrix components in the specimen system of axes, 1, 2, 3, of the kth lamina and are obtained from the values in the axes related to the fibre direction, x, y, z, by using the appropriate geometric transformation. For a beam specimen, the stresses σ_2 and σ_6 (transverse and interlaminar shear) can be neglected in comparison with σ_1.

These stresses can then be converted from the specimen axes to the fibre directions by geometric transformation. It is then possible to calculate the stresses in the fibre direction, σ_x (i.e. σ_L), normal to it, σ_y (i.e. σ_T) and the shear components, σ_{xy} (i.e. σ_{LT}). The total energy stored in the x (or L) direction, say Z_L, can then be calculated, and the energy dissipation in this layer and in this direction is then given by

$$\Delta Z_L = \psi_L Z_L \tag{22}$$

For the beam, the overall specific damping capacity ψ_{ov} is then given by

$$\psi_{ov} = \frac{\Sigma \Delta Z}{\Sigma Z} \tag{23}$$

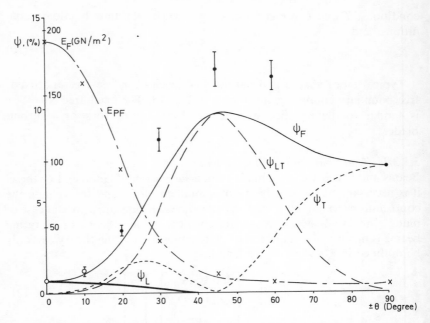

FIG. 14. Variation of flexural Young's modulus E_F and damping ψ_F with ply angle $\pm\theta$ for HM-S carbon fibre in DX209 epoxy resin, $v_f = 0.5$. Experimental points: ×, E_F; ●, ψ_F; ○, ψ_F *in vacuo*. Pure flexure prediction: ———, E_{PF}; ———, total ψ_F.

If the elastic moduli and damping coefficients are known for unidirectional material, it is possible to calculate the overall damping of a beam.

Whereas specimens with all layers at θ will twist as they are bent, the twisting can be internally restrained by using several layers at $\pm\theta$. The damping contributions can again be assessed and the measured values accounted for, as shown by Adams and Bacon [6]. Figure 14 shows theoretical predictions and experimental measurements for the modulus and damping of a series of CFRP beams made with ten layers (of high modulus carbon fibres in epoxy resin) alternately at $\pm\theta$. Note that the modulus is higher than that of the off-axis specimens owing to the internal restraint, while the damping is generally lower.

More generally laminated composites are commonly used in practice. Fortunately, the same method can be used as shown above for predicting the damping, and Fig. 15 shows the excellent agreement between theory and experiment for the variation of damping with θ of a symmetrical, E-

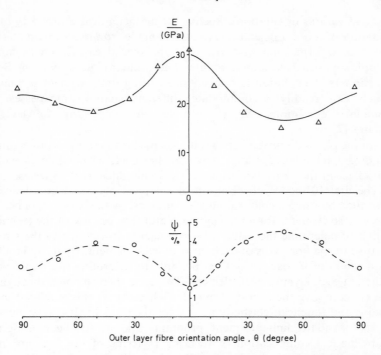

FIG. 15. Variation of flexural modulus E and damping ψ with outer layer fibre orientation angle θ for specimens cut at angle θ to 0° layer of $(0°, 90°, 45°, -45°)_s$ plate in glass/DX210. ——, theoretical E, $\triangle$; experimental E; ---, theoretical ψ; $\bigcirc$, experimental ψ.

glass fibre reinforced epoxy plate. These tests were carried out on beam specimens cut at angles from $-90°$ to $+90°$ relative to the fibre direction in the outer layer of this $(0°, 90°, 45°, -45°)_s$ plate.

5.3.2.3. *Laminated plates*

Fibre-reinforced plates of various shapes with different boundary conditions (free, clamped, hinged) commonly occur in practice. Designers need to be able to predict the stiffness parameters and damping values of such plates for conditions such as aeroelasticity, acoustic fatigue and so on. Much attention has been devoted to the stiffness predictions, but very little to damping. Our interest here is mainly in damping and the development of a suitable mathematical model which can be used to predict the damping values of plates laminated from fibres of various types at various orientations. Such is the mathematical complexity of the equation of motion of plates (even those made from isotropic materials) that closed-form solutions exist only for special cases such as hinged (simple supported) rectangular plates, and circular plates (involving Bessel functions). The solution is best obtained using finite element techniques which can readily accommodate different shapes, thicknesses and boundary conditions. Some examples are given by Cawley and Adams [7].

All the plates discussed here are mid-plane symmetric so as to eliminate bending–stretching coupling. It would, however, be possible to include this effect in the analysis if asymmetrical laminates were to be used.

The first 10 modes of vibration of a typical plate can be adequately described by using a coarse finite element mesh with six elements per side ($6 \times 6 = 36$ elements for a rectangular plate). The essence of the technique is first to determine the values of strain energy stored due to the stresses relative to the fibre axes of each layer of each element. By using modulus parameters of $0°$ bars, it is then possible to determine the total energy stored in each layer of each element. These are then summed through the thickness to give the energy stored in each element: this is related to the strains and the mean elasticity matrix for the element. It is then possible to use standard finite element programs and avoid the considerable mathematical complication of working in terms of the standard plate equations. This approach provides the stiffness of the plate, the maximum strain energy U stored in any given mode of vibration, the natural frequencies, and the mode shape. The energy dissipated in an element of unit width and length situated in the kth layer can now be determined. This is done by transforming the stresses and strains to the fibre directions and using the damping properties of $0°$ bars. The energy dissipated in the element in the kth layer is integrated over the whole area of the plate, and the contributions of each layer are summed to give ΔU, the total energy dissipated in the plate. The overall specific damping capacity ψ_{ov} is then given by $\psi_{ov} = \Delta U / U$. Alternatively, the damping can first be summed through the thickness of the damped element to give a damped element stiffness matrix. This can then be treated by standard finite element techniques.

A detailed analysis of the mathematics for the vibration of damped plates is given by Lin, Ni and Adams [8]. Figure 16 shows for the first six modes the theoretical prediction and the experimental results of carbon and glass FRP $0°/90°$ plates. On the whole, there is good agreement between the predicted and measured values. The discrepancies in natural

(a)

No.	Freq. (Hz)	Mode shape	S.D.C.(%)
1	58.10 (68.88)	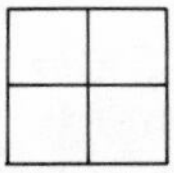	7.80 (6.65)
2	213.31 (218.9)	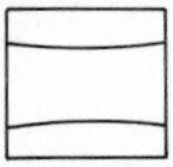	0.91 (1.05)
3	243.47 (251.2)	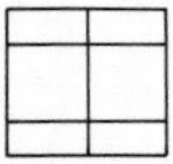	2.50 (2.6)
4	302.51 (305.4)	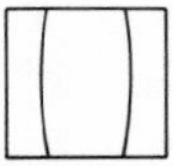	0.60 (0.92)
5	324.16 (323.5)	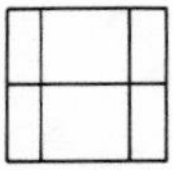	1.51 (1.7)
6	441.62 (452.5)	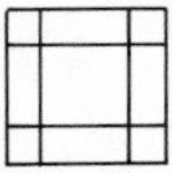	2.74 (3.0)

Outer layer Fibre direction →

(b)

No.	Freq. (Hz)	Mode shape	S.D.C.(%)
1	66.42 (62.2)	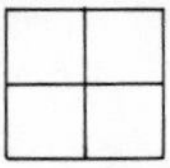	7.16 (6.7)
2	131.62 (131.4)		2.47 (2.8)
3	164.46 (159.2)	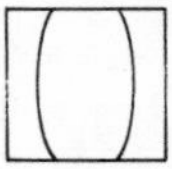	1.62 (1.9)
4	189.79 (180.5)	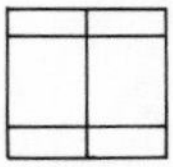	4.87 (4.9)
5	208.87 (200.05)	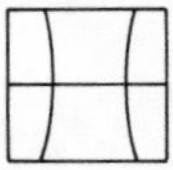	3.73 (3.2)
6	347.16 (326.7)	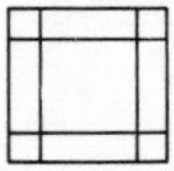	5.09 (5.8)

Outer layer Fibre direction →

FIG. 16. Natural frequencies and damping of various modes of 8-layer (0°, 90°, 0°, 90°, 90°, 0°, 90°, 0°) carbon FRP (a) and glass FRP (b) plates. Experimental values in brackets.

frequencies are less than 10%, and the values of specific damping capacity are very close. It can be said that the more the twisting, the higher the damping. For instance, for the 8-layer crossply (0°/90°) GFRP plate the two beam-type modes, numbers 2 and 3, appear similar, but the relationship of the nodal lines to the outer fibre direction means that the higher mode has much less damping than the lower one. The other modes of vibration of this plate all involve much more plate twisting and hence matrix shear than do modes 2 and 3, and so the damping is higher.

5.3.3. Effect of cracks and other damage on the damping and dynamic modulus

Consider the cracks shown in Fig. 17. Depending on whether the crack is normally open or closed, and in which direction the loads are applied, the effect of the crack will be markedly different. For instance, when shear

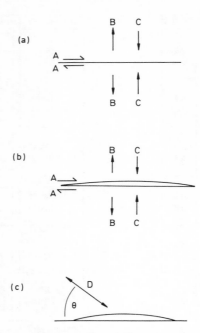

FIG. 17. Cracks under static and dynamic loads: (a) zero volume crack, both sides touching; (b) normally open crack or void; (c) part open, part closed crack.

loading, A, is applied to crack (a), the relative motion between the sides of the crack will result in energy dissipation through friction, while this will not happen with the open crack (b). However, both will exhibit a locally reduced stiffness, since even the frictional forces in crack (a) will provide small restraint compared with that of an otherwise intact structure. When tensile loading, B, is applied to either crack, little energy dissipation will occur, but there will be a local reduction in stiffness. Compressive loading, C, will close crack (a) and tend to close (b), but whether closure occurs in the latter case is dependent on the load and the initial crack size.

In general, cracks tend to be open in the middle and closed at the ends, as shown in Fig. 17c. Also, the loading, D, may be at some angle θ to the plane of the crack, resulting in both shear and tensile or compressive forces. Sliding friction will occur at the crack ends and larger normal motion at the middle, leading to both a local reduction in stiffness and a local dissipation of energy.

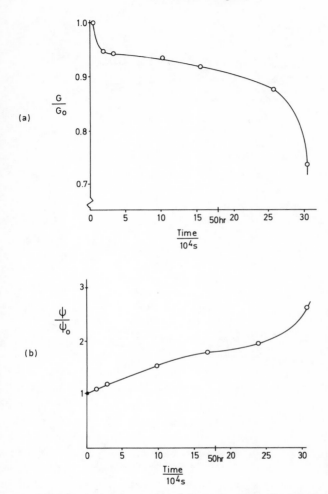

FIG. 18. Effect on unidirectionally reinforced GFRP of exposure to steam: (a) change of shear modulus ratio with time; (b) change of damping modulus ratio with time (from Adams [9]).

There are, of course, many forms of damage other than simple cracks. For instance, the material may have been subject to microplastic strain by creep or fatigue. Alternatively, some form of environmental attack may have taken place, such as by water or solvents in polymers, or by hydrogen embrittlement in metals, or by nuclear radiation in any material. Under

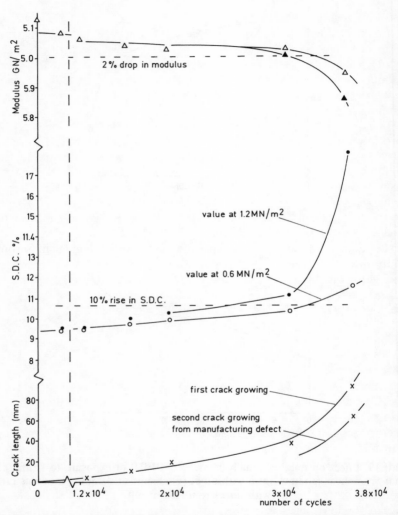

FIG. 19. Variation of damping, dynamic modulus and crack length with fatigue history of a longitudinally reinforced glass fibre in epoxy resin specimen (from Adams et al. [11]).

these circumstances, the defective zone may be local or general in the component.

In composites, a common cause of trouble is attack by aggressive environments, such as steam or solvents. For example, when cylindrical unidirectional specimens of CFRP or GFRP are subjected to steam or boiling water, the matrix is softened [9]. Figure 18 shows that the shear modulus reduces with exposure, while the damping increases. Both of these parameters are expressed as a proportion of the modulus and damping of the virgin material.

Damage induced by torsional fatigue occurs in the matrix and at the fibre–matrix interface, but rarely within the fibres. Adams and co-workers [10–12] showed that the damping increases with fatigue history while the modulus decreases, as shown in Fig. 19. Work in the same laboratory showed that static loading to produce longitudinal shear cracks has a similar effect on damping and modulus. Guild and Adams [13] have shown that cracks through the fibres in a 0° GFRP beam can also be detected by damping measurements.

Thus, it is probable that most types of stress-induced damage in fibrous composites are manifest by local or global (depending on the nature of the damage) reductions in stiffness and increases in damping. A reduction in stiffness implies a reduction in the structural natural frequencies, and hence there is the possibility of using natural frequency measurements as a non-destructive test. Similarly, the damping of the different modes of vibration may be measured.

5.3.4. Effect of dimensional and similar changes

The expressions for the flexural natural frequencies of a beam, shown on Fig. 13, indicate that natural frequencies are very sensitive to dimensional change, and so natural frequency measurements may be used to check that a component is within the prescribed tolerances.

In fibrous composites, the omission or incorrect orientation of a ply layer may be difficult to ascertain without destroying the structure. However, such a defective structure will have changes in its natural frequencies compared with a perfect one and a vibration technique may therefore be used to obtain a quick indication of the structural condition. In structures such as filament wound tubes, incorrect fibre winding, the omission of a tape helix and incorrect cure also affect natural frequencies and so may be detected.

5.4. METHODS OF MEASURING THE NATURAL FREQUENCIES AND DAMPING CHARACTERISTICS OF SPECIMENS AND STRUCTURES

Natural frequencies and damping values may be determined by either steady state or transient test techniques. The advent of cheap and powerful microprocessors has led to a change from steady state to transient testing in recent years, and the resultant decrease in testing time has made vibration methods more attractive for NDT. Only the transient testing techniques are discussed here. Details of the steady state methods may be found in, for example, Adams and Cawley [14].

One of the simplest means of measuring the damping of a system with a single degree of freedom is to displace it, release it from rest and measure the rate at which the resulting vibration dies away. This is frequently expressed in terms of the logarithmic decrement, δ, defined as

$$\delta = \log_e \left(\frac{u_n}{u_{n+1}} \right) \qquad (24)$$

where u_n is the amplitude of the nth cycle and u_{n+1} is the amplitude of the next cycle. It may readily be shown that the specific damping capacity is related to the logarithmic decrement by

$$\psi = 2\delta \qquad (25)$$

The technique as described above is not easy to apply to continuous structures because it is very difficult to displace the structure in the mode shape of a particular mode of vibration. The obvious means of applying transient excitation to a structure is to tap it with a hammer. When a structure is given an impulse, it will oscillate in all its modes of vibration, the relative strengths of the different modes being dependent on the nature and position of the impulse. Thus the structural response is a function of the natural frequencies and damping of all the modes of the system.

In recent years, fast analogue–digital converters and digital transient capture equipment have become available which enable a transient signal to be stored digitally in a computer memory. These data can be processed to yield the logarithmic decrements of each mode, but a more powerful technique is available. Modern digital spectrum analysers contain a fast Fourier transform program by which the amplitude–time data can be converted to amplitude–frequency data. A full description of the mathematics of the method is beyond the scope of this chapter, but the reader

FIG. 20. Time record of acceleration response of rectangular aluminium plate to impact. The spectrum derived from this record is shown in Fig. 11 (from Cawley and Adams [5]).

who seeks more information is referred to the excellent book by Randall [15].

Using this technique, the time record of the response of a structure to an impulse such as that shown in Fig. 20 may be converted to the corresponding frequency spectrum shown in Fig. 11. The natural frequencies of the test structure are readily identified from the maxima of the spectrum. The accuracy with which these frequencies can be defined is a function of the resolution of the frequency analyser.

By using a further technique called ZOOM analysis (also referred to as band selectable Fourier analysis), it is possible to concentrate the frequency data over a narrow range in the region of a resonant peak. It is then possible to determine the resonant frequency more accurately and also to determine the damping by the half-power bandwidth technique (provided the damping is linear).

The measurement of structural natural frequencies by this technique is extremely quick, the testing time being about 1 second, and the apparatus is very simple and easy to set up. This makes the method very attractive for non-destructive testing. A block diagram of the testing configuration used to check the natural frequencies of filament-wound CFRP tubes is shown in Fig. 21. Excitation was provided simply by tapping the top of the tube with a coin, the resulting vibration being detected by an accelerometer which was attached to the tube with wax. The signal from the accelerometer was passed to the spectrum analyser via a charge amplifier. Further savings in set-up time could be achieved by substituting a microphone for the accelerometer.

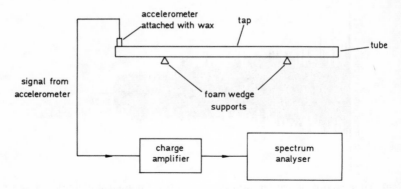

FIG. 21. Schematic diagram of apparatus used to check natural frequencies of filament-wound CFRP tubes (from Cawley, Woolfrey and Adams [16]).

The calculation of modal damping values from transient test data takes a little longer than the determination of natural frequencies, and more care is required in setting up the supports for the structure, but the use of impulse excitation and digital signal processing is the most attractive damping measurement technique for non-destructive testing purposes. With modern analysers, the same experimental data can readily be processed to yield the natural frequencies and corresponding damping factors for several modes.

5.5. GLOBAL METHODS

5.5.1. Natural frequency measurements
5.5.1.1. *Production quality control*

Since natural frequencies are very sensitive to dimensions, their use for the detection of small cracks at the production stage is limited to components which are produced to strict dimensional tolerances. More generalised defects may be found in many components and structures. Frequency measurements can, of course, be used to check whether the dimensions are within the specification.

Cawley *et al.* [16] have investigated the use of natural frequency measurements for the production quality control of fibre composites. Flexural vibration tests were carried out on a series of filament-wound carbon fibre-reinforced plastic (CFRP) tubes, several of which had deliberately built-in defects. Some of the defective tubes had incorrect

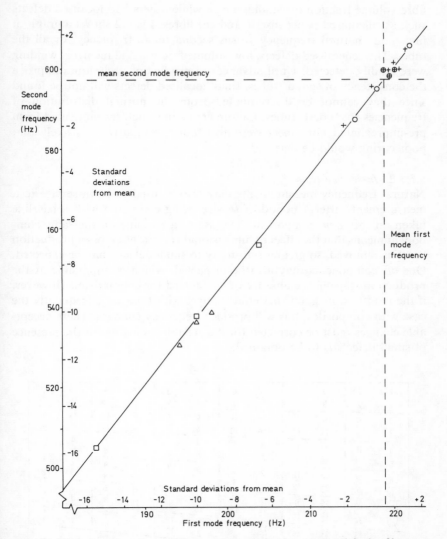

Fig. 22. First and second mode flexural frequencies for good and faulty filament-wound CFRP tubes (from Cawley, Woolfrey and Adams [16]). +, good tube ($\pm 45°$); ○, tube with siliconised paper inserts ($\pm 45°$); ⊕, tube with cut fibres ($\pm 45°$); △, misaligned tube ($\pm 50°$); □, tube with low fibre volume fraction ($\pm 45°$).

fibre volume fraction or winding angle while others had localised defects such as siliconised paper inserts and cut fibres. Fig. 22 shows a graph of first mode natural frequency versus second mode frequency for all the tubes. The generalised defects, low volume fraction and incorrect winding were readily detected. Until advances in production techniques improve the consistency of 'good' tubes, small localised defects will not be found since they cannot be discriminated from the normal distribution of frequencies for 'good' tubes. Lay-up errors in structures fabricated from pre-impregnated fibre sheet were also found. Similarly, it is likely that poor curing would be detected.

5.5.1.2. *In-service tests*

Natural frequency measurements may also be used as an in-service test, measurements after a period of service being compared with a baseline taken on the *same* component. The use of a baseline on the same component means that the effect of dimensional variations across a production batch is removed, so greater sensitivity to small defects may be expected. One-off components may also be inspected, which is impossible at the production stage in the absence of a standard for comparison. However, if the modulus of 'good' material changes with time, as is frequently the case with composites, this will produce frequency changes. These acceptable changes must be corrected for if a reliable indication of the presence of small defects is to be obtained.

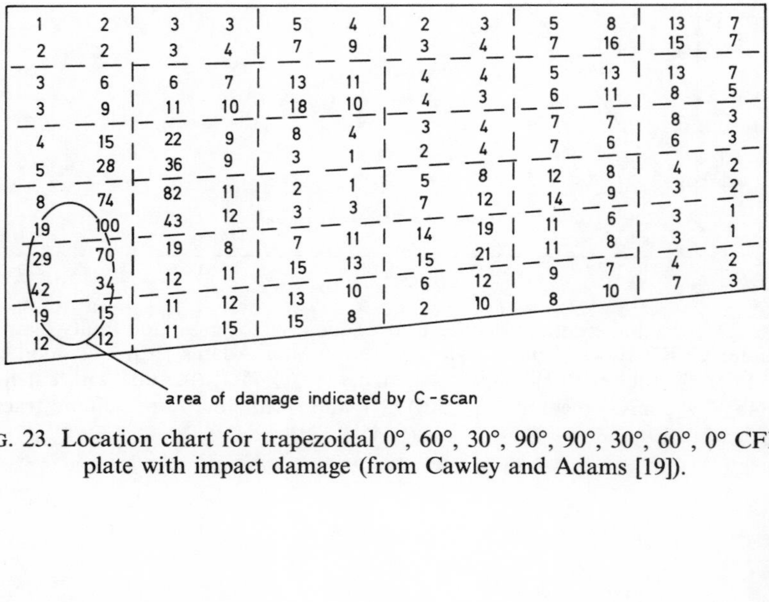

FIG. 23. Location chart for trapezoidal 0°, 60°, 30°, 90°, 90°, 30°, 60°, 0° CFRP plate with impact damage (from Cawley and Adams [19]).

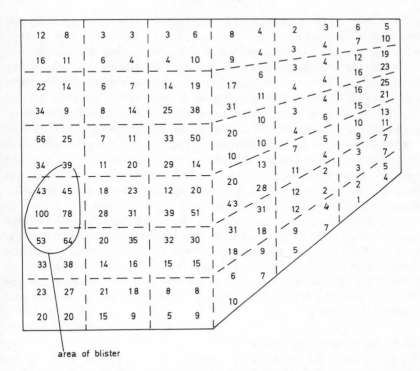

FIG. 24. Location chart for CFRP-skinned honeycomb sandwich panel damaged by local heating (from Cawley and Adams [20]).

Natural frequency measurements may also be used to locate the position of damage within a structure. The stress distribution through a vibrating structure is non-uniform and is different for each natural frequency (mode). This means that any localised damage affects each mode differently, depending on the particular location of the damage. The damage may be modelled as a local decrease in the stiffness of the structure; so, if it is situated at a point of zero stress in a given mode, it will have no effect on the natural frequency of that mode. On the other hand, if it is at a point of maximum stress, it will have the greatest effect. By comparing the frequency changes in different modes of vibration, the position of the damage may be determined.

The authors have carried out extensive tests on the method [17–20]. Over forty tests were carried out on one- and two-dimensional structures with a variety of forms of damage including holes, saw cuts, fatigue cracks, crushing, impact damage and local heating. All these forms of damage were successfully detected and located by comparing the natural frequencies measured before and after damage. An indication of the severity of the damage was also obtained. The results were presented in the form of a damage location chart which gives a measure of the probability of damage being at a given location. A value of 100 indicates the most probable site, lower values representing decreasing probabilities.

One of the tests was carried out on a trapezoidal CFRP plate which was subjected to impact damage. An ultrasonic C-scan of the plate revealed extensive delaminations, but there was little visible damage. The location chart for this case, given in Fig. 23, shows that the damage was successfully detected and located by natural frequency measurements.

Two tests were carried out on honeycomb constructions. One of these was a $610 \times 520 \times 10$ mm panel with CFRP facings of the type which is used for floor panelling in aircraft. The panel was made asymmetrical by removing one corner and was damaged by heating one face with a gas flame. This produced a large blister on this face but the damage was invisible from the remote side. Figure 24 shows that the damage was successfully located.

5.5.2. Damping measurements

Structural damping is much less sensitive to dimensions than are the corresponding natural frequencies. Damping measurements may therefore be used as a production quality control test, even on components which are not produced to tight dimensional tolerances. They may also be used as an in-service test in the same manner as natural frequency measurements. However, although damping tends to be more sensitive to damage than natural frequencies, it is much more difficult to measure accurately. In particular, care must be taken to minimise extraneous damping from, for example, the system supporting the component during the test.

Another difficulty with the use of damping measurements for the detection of damage in fibre composite components is that the level of damping in a 'good' component tends to be higher than that in a metal component of similar construction. Therefore a localised defect will produce a smaller percentage change in the damping. This, coupled with the difficulty of accurate measurements, means that damping measurements are unlikely to find widespread use in the non-destructive testing of fibre composites.

5.6. LOCAL MEASUREMENTS

5.6.1. Excitation at a single point

Techniques in this category involve vibrating the test structure (usually at resonance) by applying an exciting force at a single point and measuring a local property of the structure in the particular mode of vibration at all the points of interest. If these measurements can be carried out by a scanning system, the test is potentially quick to carry out and, because local properties are being measured, it may be more sensitive than the global methods described earlier.

5.6.1.1. *Vibrothermography*
The presence of damage in fibre-reinforced plastics results in the formation of cracks and crazes. When cyclic stresses are applied to a damaged composite material, relative motion (damping) takes place between the sides of the assorted cracks, resulting in the generation of heat. The change in the overall level of damping in a structure is small for many forms of serious but localised damage, while the damping in this small local area may be large.

The local temperature rise caused by the cyclic stressing can be measured by a variety of means, such as temperature-sensitive paints or crystals, but is usually measured by a scanning infra-red camera. A camera such as the Barr and Stroud, or one of the Agavision systems, can provide a graded black/grey/white or false-colour representation of the surface temperatures. Errors can occur owing to variations in surface thermal emissivity, since perfect radiation is only achieved with an ideal black-body radiator.

Various authors have used cyclic stresses together with an infra-red camera, usually referred to as vibrothermography, for non-destructively examining structures. Usually, it is necessary that the structure has low thermal diffusivity to prevent the rapid conduction of heat from the damaged area. It is therefore of least application to metals (unless the dissipation rate is large) and of greatest application to materials such as GFRP. Carbon-based composites have markedly higher thermal conductivity than glass-based ones, and are not so easy to test.

Most authors (e.g. Reifsnider and Williams [21], Nevadunsky *et al.* [22]) have applied the cyclic stresses using servo-hydraulic testing machines, but some such as Reifsnider and Stinchcomb [23], Pye and Adams [24] and Russell and Henneke [25] have used resonant vibration in the audible frequency range, while yet others have successfully used axial stresses applied with an ultrasonic vibrator (Henneke and Jones [26]).

Pye and Adams [24] used resonant vibration to detect shear cracks in unidirectional fibrous composites consisting of glass or carbon fibres in an epoxy resin matrix. Cylindrical specimens were vibrated in torsion by attaching one end to a rigid abutment and the other end to an inertia which was then excited using coils and magnets. They carried out an analysis to determine the surface temperature rises that could be expected from such a test, and confirmed some of the results experimentally. Pye and Adams [27] also considered the effect of the cyclic stress and frequency of vibration on the temperature rise. They showed that, if other parameters remain constant, the temperature difference, ΔT, between any two points of the structure is proportional to $\sigma^n f$ where σ is the peak cyclic stress, f is the frequency of vibration, and the exponent n is a function of the damping–stress relationship in the region of the crack. In practice, other parameters, such as the specimen stiffness, also change, but the most significant is the increase in damping with stress for a damaged structure.

In Section 5.2.1, it was shown that $\psi = \Delta W/W$, where ΔW is the energy dissipated, which appears as heat, and W is the maximum cyclic strain energy which is $\sigma^2/2E$ per unit volume. If ψ is independent of cyclic stress, then ΔW is proportional to σ^2. If ψ increases with cyclic stress, as is usually the case, then the exponent n is > 2.

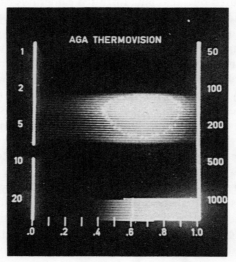

FIG. 25. Thermograph for GFRP tube.

Using resonant vibration (hence the definition 'resonant vibrothermography), it is possible to use higher cyclic frequencies than can be obtained using servo-hydraulic machinery. This increases the sensitivity of the test, thus allowing easier detection of the damage or, alternatively, a reduction in the cyclic stress used. McLaughlin *et al.* [28] found that with CFRP it was difficult to detect damage at low frequencies (1–5 Hz), and suggested the audio frequency range as being the most suitable. Higher frequencies (> 15 kHz) are unsuitable as the cyclic stresses tend to be too small.

Although there are problems associated with vibrothermography (notably thermal conduction, stray heat sources, temperature sensitivity and surface thermal emissivity), it is a non-contacting technique which can be used at a distance, and crack growth can be monitored as it occurs.

A thermograph taken during vibration of a $\pm 30°$ filament-wound impact-damaged GFRP tube is shown in Fig. 25.

5.6.1.2. *Local amplitude measurement*

Local damage in a structure tends to distort the vibration mode shapes. With the advent of laser holography systems, there has been an upsurge of interest in the exploitation of this effect in NDT [29]. The technique involves vibrating the test structure at resonance and producing a time-averaged hologram of the motion (Campbell and McLachlan [30]). The method has been particularly successful for detecting skin-core disbonds in honeycomb panels. However, it is essential that the system be set up in a vibration-free environment. This means that it is only suitable for specialist applications.

The use of pulsed lasers makes it possible to use the technique in the presence of background vibration (Fagot *et al.* [31]). In this case, the best results are obtained by using impulse excitation and producing the hologram from views taken a few microseconds apart as the pulse propagates through the structure. An example of the results obtained on an aluminium honeycomb panel containing two defects is shown in Fig. 26, the field observed being about 600 mm across. The technique is still under development and the equipment costs are high, but rapid inspection is possible and the method may prove cost-effective where large areas of structure are to be inspected.

A similar idea has been employed on multi-layer structures in the Soviet Union (Lange and Moskovenko [32]). The investigators excited the structure over a wide frequency band above the major structural resonances and used lycopodium powder sprinkled on the surface of the structure for

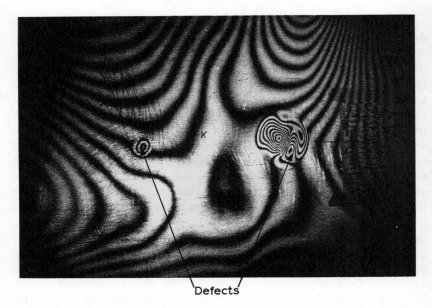

FIG. 26. Double exposure holographic view of defects in an aluminium honeycomb structure (from Fagot et al. [31]).

detection. In defective zones, the amplitude was higher due to the membrane resonance effect discussed in Section 5.6.2.3 and so the powder migrated away from these areas. The method is very quick and simple to apply but can only detect flaws to a depth of 3–5 mm. It also depends on there being a membrane resonance of the layer above the defect in the frequency range covered by the exciter.

5.6.2. Excitation at each test point

The methods described in this section all involve applying vibration excitation at each test point and measuring the input force and/or the vibration response at the same point. They are used for the detection of defects such as disbonds in adhesive joints, delaminations and voids in laminated structures and defective honeycomb constructions. These are all 'planar' defects which result in one or more layers of the construction being separated from the base layer(s). The methods are not suitable for detecting transverse cracks (i.e., those running in a direction normal to the surface).

The main advantage of these techniques over the higher frequency ultrasonic methods is that no coupling fluid is required between the transducer and the structure, which results in more rapid and convenient testing. They are also easier to apply to honeycomb constructions.

5.6.2.1. *The coin-tap test*

As discussed earlier, the coin-tap test is one of the oldest methods of non-destructive inspection. It has been used regularly in the inspection of laminates and honeycomb constructions. Indeed, Hagemaier and Fassbender [33] found that it could detect more types of defect in honeycomb constructions than any other technique except neutron radiography. Until recently, however, the technique has remained largely subjective and there has been considerable uncertainty about the physical principles behind it.

It should be stressed again that this test is quite different from the wheel-tap test discussed in Section 5.5, though the testing technique and subjective interpretation of the sound produced is similar in both cases. The wheel-tap test is a global test which investigates the whole test component from a tap applied at a single point, the difference between sound and defective components being detected from changes in the natural frequencies and damping. The coin-tap test will only find defects in the region of the tap, so it is necessary to tap each part of the structure under investigation.

The sound produced when a structure is tapped is mainly at the frequencies of the major structural modes of vibration. These modes are structural properties which are independent of the position of excitation. Therefore, if the same impulse is applied to a good area and to an adjacent defective area, the sound produced must be very similar. The difference in the sound produced when good and defective areas are tapped must therefore be due to a change in the force input.

When a structure is struck with a hammer, the characteristics of the impact are dependent on the local impedance of the structure and on the hammer used. Damage such as an adhesive disbond results in a local decrease in structural stiffness and hence a change in the nature of the impact. The time history of the force applied by the hammer during the impact may be measured by incorporating a force transducer in the hammer. This technique is commonly used in structural dynamic testing (Peterson and Klosterman [34]). Typical force–time histories from taps on sound and debonded areas of an adhesively bonded structure are shown in Fig. 27. The impact on the sound structure is more intense and of shorter duration than that on the damaged area, the impact duration

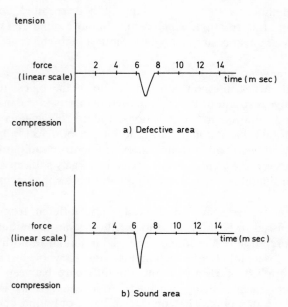

FIG. 27. Force–time records for impacts on defective and sound regions of adhesively bonded structure (from Adams and Cawley [46]).

on the sound structure being approximately 1 ms. The difference between the two impulses is more readily quantified if the frequency content of the force pulses is determined. This is achieved by carrying out a Fourier transform of the force–time records. The spectra derived from the force–time histories shown in Fig. 27 are given in Fig. 28. The impact on the damaged area has more energy at low frequencies but the energy content falls off rapidly with increasing frequency, while the impact on the sound area has a much lower rate of decrease of energy with frequency. This means that the impact on the defective area will not excite the higher structural modes as strongly as the impact on the good zone. The sound produced will therefore be at a lower frequency and the structure will sound 'dead'.

The testing technique therefore involves tapping the area of the structure to be tested with an automatic, instrumented hammer designed to give a single, reproducible impact and comparing the frequency spectrum of the impulse with that of an impulse from the same hammer on an area of the same type of structure which is known to be sound. Data from a sound

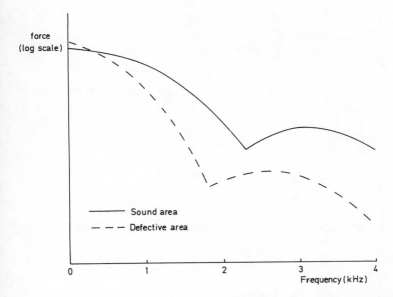

FIG. 28. Spectra of time records shown in Fig. 27 (from Adams and Cawley [46]).

structure would be stored in the testing instrument so that the instrument could carry out the comparison and give an immediate indication of the integrity of the area under test. This development is discussed in more detail by Adams *et al.* [35].

5.6.2.2. *The impedance method*

The impedance method of non-destructive inspection has been used in the Soviet Union for almost thirty years (Lange and Moskovenko [32]) and its use in the West has now increased with the marketing of the 'acoustic flaw detector', which is based on the Soviet design, by Inspection Instruments Limited. The technique uses measurements of the point impedance, Z, of a structure defined as

$$Z = \frac{P}{v} \tag{26}$$

where P is the harmonic force input to the structure and v is the resultant velocity of the structure at the same point. The measurements are carried out at a single frequency, typically between 1 and 10 kHz [32]. There has been considerable interest in the technique, particularly from the aero-

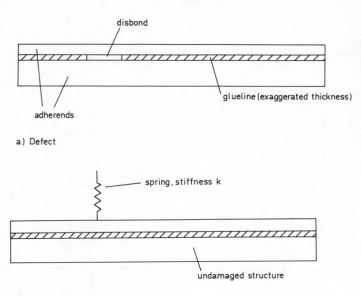

Fig. 29. Typical defect and mathematical model (from Cawley [36]).

space industry, but uncertainty about which physical parameters are measured, and the relationship between these and the presence of defects, has limited its application.

The impedance method seeks to detect areas of a structure where one or more layers are separated from the base layer(s) as shown in Fig. 29a. A localised defect of this type has little effect on the overall dynamic properties of the structure. It has been shown (Cawley and Adams [18, 20]) that the changes in mode shapes and structural natural frequencies produced by localised defects, while measurable, are in general small. Clearly, however, the local stiffness of the structure is significantly reduced. The defect may therefore be modelled as a spring, below which is the rest of the structure whose properties are unaltered, as shown in Fig. 29b. The spring stiffness is given by the static stiffness of the layer(s) above the defect supported around the edges of the defect. The boundary condition is likely to be intermediate between simply supported and clamped.

The plate formed by the layer(s) above the defect can resonate. As the first resonant frequency, which is that of the membrane resonance discussed in Section 5.6.2.3, is approached, the effective stiffness of the plate

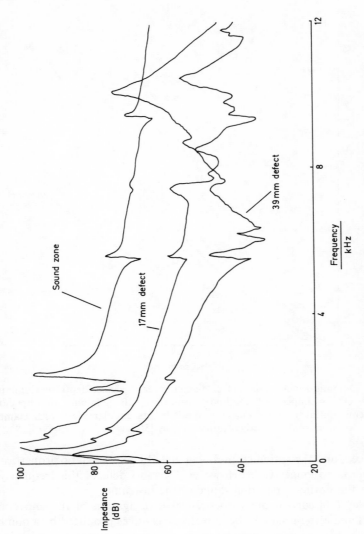

FIG. 30. Measured impedance–frequency curves for thick beam (after Cawley [36]).

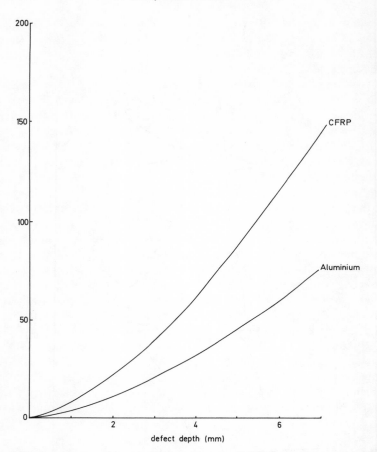

FIG. 31. Minimum detectable defect diameter versus defect depth in aluminium and CFRP for impedance method, assuming 3 dB reliability in impedance measurement. Layer above defect is modelled as circular plate with clamped edges (after Cawley [36]).

is reduced, and the spring stiffness calculated from the static properties of the layer is no longer valid. However, for most defects, this frequency is outside the normal operating range of the instrument.

Cawley [36] carried out a theoretical investigation of the impedance method and also measured the impedance changes produced by a number of defects. Fig. 30 shows impedance–frequency curves taken on a laminate comprising an aluminium sheet 45 mm wide and 3.3 mm deep bonded to a 30 mm deep steel beam. Two areas of the sheet, one 39 mm long and the other 17 mm long, were not coated with adhesive and so formed disbonds. Fig. 30 shows that the defects can clearly be detected from impedance measurements.

These measurements were carried out using an impedance head (Ewins [37] and Brownjohn et al. [38]) cemented to the structure. For practical non-destructive testing, a dry point contact between the transducer and the structure is used. This contact has a finite stiffness (Lange and Teumin [39]), so the spring stiffness in the model becomes the defect stiffness in series with the contact stiffness. The spring stiffness over a good area of the structure is the contact stiffness alone. It is important to keep the contact stiffness as high as possible as this is the major limitation on the sensitivity of the technique. The method is also most sensitive to defects between a thin top layer and a stiff base structure. Fig. 31 shows the variation of minimum detectable defect diameter with depth for aluminium and carbon fibre-reinforced plastic structures, assuming a stiff base structure and a 3 dB reliability in impedance measurement (i.e., the impedance in a defective zone must be at least 3 dB lower than that in a sound zone if the defect is to be reliably detected). Typical values of contact stiffness were assumed, the technique being less sensitive for the CFRP facing because of a lower contact stiffness.

Problems have arisen with the systems currently employed both in the Soviet Union and the West because the transducers used, unlike the impedance head used in the tests reported by Cawley [36], give an output which is a highly non-linear function of the structural impedance and the simple physical interpretation which can be made of the results shown in Fig. 30 is lost.

5.6.2.3. *Membrane resonance methods*

It was shown in the previous section that the layer of material above a disbond or delamination may be modelled as a plate which is restrained around the defect edges. This plate can resonate, the first mode being the membrane resonance in which the motion is similar to that of a diaphragm.

As the frequency of vibration excitation approaches the membrane resonant frequency, the impedance of the system is reduced and the response amplitude obtained for a given force input increases. Therefore, at frequencies around the membrane resonance of the layer above a defect, the vibration response for a given force will be greater than that obtained in a sound region of the structure. Thus a non-destructive test may be based on response measurements alone, assuming the input force to be approximately constant. Since the resonant amplification tends to be large (greater than a factor of 10, which is 20 dB), the controls on the size of the force need not be too sophisticated. The test is therefore potentially simpler than the impedance method discussed in the previous section which relies on measuring impedance changes of the order of 3 dB.

The technique is most sensitive if excitation is applied at the membrane resonant frequency. Since this frequency is dependent on defect size and depth, a broad band of frequencies must be covered. This was achieved in the Fokker bond tester, type 1, by using white noise excitation in a band from 0.5 to 10 kHz (Schliekelmann [40]). Transmitting and receiving piezoelectric probes were used and a meter displayed the ratio between the transmitted and received energy. It should be noted that this instrument is quite different from the widely used Fokker bond tester type 2 which operates at much higher ultrasonic frequencies (200 kHz upwards) and requires a couplant between the transducer and the structure.

It is also possible to use the technique with a single excitation frequency, though the sensitivity to flaws whose membrane resonant frequencies are far removed from the excitation frequency is considerably reduced. The 'harmonic bond tester' developed by Boeing and produced by the Shurtronics Corporation excites the laminate by induced eddy currents and measures the response with a microphone within the eddy current coil (Phelan [41]). The interaction between the original and induced fields produces vibration excitation at double the frequency applied to the coil. The harmonic bond tester uses a 15 kHz oscillator and so produces vibration excitation at 30 kHz. The eddy current excitation system cannot be used with nonconducting materials, for which alternative excitation methods are employed. Botsco [42] discusses a similar system but does allow the excitation frequency to be varied to suit the application. A single excitation frequency has also been used in the 'acoustic amplitude' method developed by Lange [43].

All the instruments discussed in this section are dependent on the membrane resonant frequency of the layer above the defect being close to the operating frequency range of the transducer. The membrane resonant frequency of a circular plate clamped around its edges is given by

$$f = \frac{0.47h}{a^2} \sqrt{\left(\frac{E}{\rho(1-\mu^2)}\right)} \qquad (27)$$

where h is the thickness of the layer above the defect, a is the defect radius and E, ρ and μ are the material modulus, density and Poisson's ratio respectively. Thus a defect 1 mm deep and of 5 mm radius in cross-plied

CFRP has a membrane resonance at approximately 90 kHz. It is unlikely that any of the membrane resonance systems discussed above would detect this defect.

5.6.2.4. *The acoustic spectral method*
Lange [44] has developed the 'Acoustic Spectral Flaw Detector' which was designed as an automated version of the coin-tap test discussed in Section 5.6.2.1. The test area is tapped by an electromagnetically operated hammer and the resultant vibration is detected either by a microphone or by a piezoelectric transducer. The response signal is passed through a parallel set of filters to produce a response spectrum. This spectrum is then compared with that of an area which is known to be good.

The authors' work on the coin-tap test discussed in Section 5.6.2.1 showed that the presence of a delamination or similar defect reduces the high frequency energy present in the force pulse. This tends to reduce the amplitude of the response at high frequencies. However, Lange [44] and Lange and Moskovenko [32] report that the high frequency response is frequently increased by the presence of a defect. This is because the membrane resonance of the layer above the defect falls within the frequency range investigated. Later work by Lange and Ustinov [45] has shown that the presence of defects can either increase or decrease the response amplitude in different parts of the spectrum. The method therefore uses a combination of the change in input force due to damage and the membrane resonance effect discussed in Section 5.6.2.3. There is a slight possibility that the effects might cancel out. These difficulties are avoided if the spectrum of the input force rather than the response is monitored, as in the method proposed by Adams *et al.* [35].

5.6.2.5. *Velocimetric methods*
A number of techniques utilising the effect of flaws on the wave speed and wave path length between transmitting and receiving piezoelectric transducers have been developed in the Soviet Union (Lange and Moskovenko [32], Vinogradov *et al.* [46]). These use frequencies between 25 and 60 kHz which are higher than those used by most of the tests discussed earlier. The methods are more sensitive than the impedance technique discussed in Section 5.6.2.2, but unless both sides of the component can be accessed there is a dead zone adjoining the face opposite to the test surface which constitutes 20–40% of the thickness.

The Sondicator discussed by Chapman [47] may also be termed a velocimetric technique.

5.7. DISCUSSION

The global methods based on natural frequency measurements discussed in Section 5.5 have the major advantage that the whole component can be checked by measurements made at a single point. This means that very rapid testing is possible, particularly if a tap testing technique is employed. Not surprisingly, if a single point measurement is used to infer the quality of the whole component, the sensitivity of the technique tends to be lower than that obtained by local measurements such as ultrasonic inspection.

The most attractive local measurement techniques use a single excitation point and a non-contacting, scanning measurement system. Unfortunately, the equipment required tends to be very expensive. The thermographic system for local damping measurements will cost over £25 000, and if a holographic system is used for local amplitude measurements the cost is likely to be even greater. However, in large scale applications where the techniques have the required sensitivity, the cost will be justified.

The methods discussed in Section 5.6.2 require excitation at each test point and are therefore much slower. However, they have the advantage over higher frequency ultrasonic inspection that a dry contact between the probe and the structure is satisfactory and so no coupling fluid is required. They are therefore easier to apply, particularly *in situ*, for example on an aircraft wing. The techniques are also more suited to the inspection of honeycomb constructions than the higher frequency methods.

All the methods of Section 5.6.2 are used for finding defects such as disbonds and delaminations, usually under fairly thin skins. The impedance method operates at a single excitation frequency, so the computational requirements are small. This means that the probe can be moved over the surface of the structure, giving a continuous reading. However, there are dangers in using only one frequency.

The automated coin-tap method requires a spectrum to be computed at each test point. This means that the inspection rate is approximately five to ten positions per second. However, the reliability is improved by looking at more than one frequency, and the tapping head only makes instantaneous contact with the structure, which removes the alignment and static clamping force problems which can arise with the impedance technique.

The methods based on the membrane resonance of the layer above the defect are quick, but problems arise with defects whose natural frequencies are above the frequency range of the instrument. The probability of missing defects is greatly increased if excitation is confined to a single frequency or a narrow band.

None of the methods discussed here will be universally applicable. However, the range of low frequency techniques based on local measurements can make a valuable contribution to the non-destructive testing of fibre composites. The local methods discussed in Section 5.6.2 are particularly useful with honeycomb constructions and composite materials where it is frequently necessary to detect delaminations and disbonds in a plane parallel to the structure's surface. This is the type of defect to which these local techniques are most sensitive.

5.8. CONCLUSIONS

The vibration techniques reviewed here make a very valuable contribution to the NDT of fibre-reinforced plastics. The 'global' methods involving measurements of natural frequencies (and possibly damping) are extremely quick to carry out and give information about the integrity of the whole of the component from a single point test. They are not as sensitive to small localised defects as tests which involve point by point testing of the whole component, but they can give valuable information both at the production stage and in-service.

The 'local' vibration techniques are particularly useful in circumstances where the use of the coupling fluids required for ultrasonic testing is undesirable. They are also well suited to the detection of defects such as disbonds and delaminations which are a common problem with composite materials and honeycomb constructions.

The advent of microelectronics means that the cost and bulk of the equipment required to carry out vibration tests reliably and quickly is reducing rapidly. The time is therefore ripe for more attention to be paid to the opportunities which vibration techniques offer for improving inspection reliability and efficiency.

REFERENCES

1. R. D. Adams and D. G. C. Bacon, The dynamic properties of unidirectional fibre reinforced composites in flexure and torsion. *Journal of Composite Materials*, January 1973, **7**(1), 53–67.
2. D. F. Adams and D. R. Doner, Longitudinal shear loading of unidirectional composite, *Journal of Composite Materials*, January 1967, **1**(1), 4–17.

3. Z. Hashin, Complex moduli of viscoelastic composites II: Fibre reinforced materials, *International Journal of Solids and Structures*, June 1970, **6**(6), 797–807.
4. R. D. Adams and D. Short, The effect of fibre diameter on the dynamic properties of glass fibre reinforced polyester resin, *Journal of Physics D: Applied Physics*, 11 June 1973, **6**(9), 1032–1039.
5. P. Cawley and R. D. Adams, Improved frequency resolution from transient tests with short record lengths, *Journal of Sound and Vibration*, May 1979, **64**(1), 123–132.
6. R. D. Adams and D. G. C. Bacon, Effect of fibre orientation and laminate geometry on the dynamic properties of CFRP, *Journal of Composite Materials*, October 1973, **7**(4), 402–428.
7. P. Cawley and R. D. Adams, The predicted and experimental natural modes of free–free CRFP, *Journal of Composite Materials*, October 1978, **12**(4), 336–347.
8. Dx Lin, R. G. Ni and R. D. Adams, Prediction and measurement of the vibrational damping parameters of carbon and glass fibre-reinforced plastics plates, *Journal of Composite Materials*, March 1984, **18**(2), 132–152.
9. R. D. Adams, Non-destructive evaluation of fibre reinforced plastics exposed to aggressive environments, *Proc. Int. Symp. Non-destructive Control of Materials*, Bratislava, Czechoslovakia, 1976.
10. R. D. Adams, J. E. Flitcroft, N. L. Hancox and W. N. Reynolds, Effect of shear damage on the torsional behaviour of carbon fibre reinforced plastics, *Journal of Composite Materials*, January 1973, **7**(1), 68–75.
11. R. D. Adams, D. Walton, J. E. Flitcroft and D. Short, Vibration testing as an NDT test tool for composite materials, ASTM STP 580, *Composite Reliability*, August 1975, 159–175; Proc. Symposium, Las Vegas, 15–16 April 1974.
12. R. D. Adams, D. Walton and D. Short, The accumulation of damage in the torsional fatigue of glass and carbon fibre reinforced material and its detection by a vibration method, *New developments in NDT of non-metallic materials*, 2nd Int. RILEM Conf. on NDT, Constanza, Romania, 4–7 September 1974.
13. F. J. Guild and R. D. Adams, The detection of cracks in damaged composite materials, *Journal of Physics D: Applied Physics*, 14 August 1981, **14**(8), 1561–1573.
14. R. D. Adams and P. Cawley, Vibration techniques in non-destructive testing, in: R. S. Sharpe (editor), *Research Techniques in Non-destructive Testing, Volume VIII*, Academic Press, London/Orlando, 1985, pp. 303–360; ISBN 0-12-639058-4.
15. R. B. Randall, *Application of B&K equipment to frequency analysis*, 2nd edition, Bruel & Kjaer, DK2850 Naerum, Denmark, 1977; ISBN 87-87355-14-0.
16. P. Cawley, A. M. Woolfrey and R. D. Adams, Natural frequency measurements for production quality control of fibre composites, *Composites*, January 1985, **16**(1), 23–27.
17. R. D. Adams, P. Cawley, C. J. Pye and B. J. Stone, A vibration technique

for non-destructively assessing the integrity of structures, *Journal of Mechanical Engineering Science*, April 1978, **20**(2), 93–100.
18. P. Cawley and R. D. Adams, The location of defects in structures from measurements of natural frequencies, *Journal of Strain Analysis*, April 1979, **14**(2), 49–57.
19. P. Cawley and R. D. Adams, A vibration technique for non-destructive testing of fibre-composite structures, *Journal of Composite Materials*, April 1979, **13**(2), 161–175.
20. P. Cawley and R. D. Adams, Defect location in structures by a vibration technique, ASME Paper number 79-DET-46, 1979, for meeting during 10–12 September 1979.
21. K. L. Reifsnider and R. S. Williams, Determination of fatigue-related heat emission in composite materials, *Experimental Mechanics*, December 1974, **14**(12), 479–485.
22. J. J. Nevadunsky, J. J. Lucas and M. J. Salkind, Early fatigue damage detection in composite materials, *Journal of Composite Materials*, October 1975, **9**(4), 394–408.
23. K. L. Reifsnider and W. W. Stinchcomb, New methods of mechanical materials testing using thermography, *Proc. 3rd Infra-red Information Exchange*, St Louis, August 1976.
24. C. J. Pye and R. D. Adams, Detection of damage in fibre reinforced plastics using thermal fields generated during resonant vibration, *NDT International*, June 1981, **14**(3), 111–118.
25. S. S. Russell and E. G. Hennke, Dynamic effects during vibrothermographic NDE of composites, *NDT International*, February 1984, **17**(1), 19–25.
26. E. G. Henneke and T. S. Jones, Detection of damage in composite materials by vibrothermography, ASTM STP 696, *NDE and flaw criticality for composite materials*, December 1979, 83–95; Proc. Symposium, Philadelphia, 10–11 October 1978.
27. C. J. Pye and R. D. Adams, Heat emission from damaged composite materials and its use in non-destructive testing, *Journal of Physics D: Applied Physics*, 14 May 1981, **14**(5), 927–941.
28. P. V. McLaughlin, E. V. McAssey and R. C. Deitrich, Non-destructive examination of fibre composite structures by thermal field techniques, *NDT International*, April 1980, **13**(2), 56–62.
29. R. K. Erf, *Holographic non-Destructive testing*, Academic Press, New York and London, 1974.
30. J. M. Campbell and E. H. McLachlan, Holographic non-destructive testing, *British Journal of Non-Destructive Testing*, March 1979, **21**(2), 71–75.
31. H. Fagot, F. Albe, P. Smigielski and J. L. Arnaud, Holographic non-destructive testing of materials using pulsed lasers, Institut Franco-Allemand de Recherches de Saint-Louis, Report No. CO 223/80, 1980.
32. Yu. V. Lange and I. B. Moskovenko, Low frequency acoustic non-destructive test methods, *Soviet Journal of Non-destructive Testing*, September 1978, **14**(9), 788–797.
33. D. Hagemaier and R. Fassbender, Non-destructive testing of adhesive bonded structure, *SAMPE Quarterly*, July 1978, **9**(4), 36–58.

34. E. L. Peterson and A. L. Klosterman, Obtaining good results from an experimental modal survey, *Journal of the Society of Environmental Engineers*, March 1978, **17**–1(76), 3–10.
35. R. D. Adams, A. M. Allen and P. Cawley, The coin-tap test for laminated structures, *Proc. 11th World Conference on NDT*, Las Vegas, November 1985, Session II, Inspection of composites and bonded structures.
36. P. Cawley, The impedance method of non-destructive inspection, *NDT International*, April 1984, **17**(2), 59–65.
37. D. J. Ewins, Measurement and application of mechanical impedance data, II: Measurement techniques, *Journal of the Society of Environmental Engineers*, March 1976, **15**–1(68), 23–30, 33.
38. J. M. W. Brownjohn, G. H. Steel, P. Cawley and R. D. Adams, Errors in mechanical impedance data obtained with impedance heads, *Journal of Sound and Vibration*, 8 December 1980, **73**(3), 461–468.
39. Yu. V. Lange and I. I. Teumin, Dynamic flexibility of a dry point contact, *Soviet Journal of Non-destructive Testing*, 1971, **7**, 157–165.
40. R. J. Schliekelmann, Non-destructive testing of adhesive bonded metal-to-metal joints 2, *Non-destructive Testing*, June 1972, **5**(3), 144–153.
41. C. S. Phelan, Critical analysis of two stage non-destructive testing using harmonic and interfacing resonant frequency systems, *Applied Polymer Symposia* (Wiley), 1972, **19**, 423–439.
42. R. Botsco, The eddy-sonic test method, *Materials Evaluation*, 1968, **26**, 21–26.
43. Yu. V. Lange, Acoustic amplitude method of inspecting bonding in laminated structure, *Soviet Journal of Non-destructive Testing*, Jan/Feb. 1976, **12**(1), 5–11.
44. Yu. V. Lange, The acoustical spectral non-destructive testing method, *Soviet Journal of Non-destructive Testing*, March 1978, **14**(3), 193–199.
45. Yu. V. Lange and E. G. Ustinov, The AD-60S low frequency acoustic flaw detector, *Soviet Journal of Non-destructive Testing*, January 1982, **18**(1), 11–14.
46. N. V. Vinogradov, E. I. Tsorin, S. A. Filimonov, Yu. V. Lange and V. V. Murashov, Low frequency acoustic flaw detector for reinforced plastics and glued members, *Soviet Journal of Non-destructive Testing*, Jan./Feb. 1977, **13**(1), 93–96.
47. G. B. Chapman, A non-destructive method of evaluating adhesive bond strength in fibreglass reinforced plastic assemblies, ASTM STP 749, *Joining of composite materials*, September 1981, 32–60; Proc. Symposium, Minneapolis, 16 April 1980.

Chapter 6

Corona Discharge

JOHN SUMMERSCALES
Royal Naval Engineering College, Manadon, Plymouth, UK

The corona discharge NDT method operates on a simple principle. If an electric field of high intensity is imposed across a void contained in a dielectric material, then the gas in the void will be ionised and electrons are accelerated towards the void wall. The void can be detected either by the minute pulse of current which occurs in the secondary coil of the transformer, or by the radiation of electromagnetic wavelengths as a result of the collision.

Hendron *et al.* [1, 2] examined glass–epoxy composites containing artificial voids and delaminations. Voids were produced by embedding sealed pieces of glass tube in the sample during manufacture. Delaminations were simulated by embedded layers of filter paper encased in Mylar. The tests were conducted under oil to eliminate the unwanted corona around the electrodes. During the raising of the voltage, the surface of the structure was scanned using the test probe while simultaneously listening for radio noise. The secondary voltage at which the corona signal first occurred was recorded. The presence of a defect was verified by examination of the trace on an oscilloscope. The corona method was shown to require a higher voltage for the detection of defects near the centre of the sample than at the surface, and a higher voltage was needed to reveal a smaller defect. Delaminations were reliably detected but required larger voltages than voids at the same depth (Table 1).

Walker *et al.* [3] have examined NOL rings reinforced with S-994 HTS glass fibre in Epon 826 resin by the corona discharge method. Tests were carried out at up to 60 kV, and at 60 Hz, with the sample immersed in mono-iso-propyl biphenyl (MIPB). The rings had a diameter of 150 mm, a width of 6.4 mm and a thickness of 3.2 mm. Time was found to be a

TABLE 1

Results of Microwave and Corona Discharge Tests by Hendron et al. [2], reproduced from *Materials Evaluation* with the Permission of the American Society for Nondestructive Testing.

(N.A. = not applicable, N.D. = not detected, — = no data)

Defect	Dimensions (mm) (dia. × length)	Panel thickness (mm)	Defect depth (mm)	Corona (kV)	Microwave (mV)
Void	5 × 12	10	5	75	7
		20	5	53	16
		20	7	72	—
		20	8	N.D.	10
		20	12	N.A.	12
		20	14	N.A.	10
Void	2 × 12	10	3	14	8
		10	4	45	6
		10	5	74	6
		20	6	N.D.	5
		20	8	55	4
		20	9	57	—
		20	12	N.A.	6
		20	14	N.A.	5
Void	1 × 12	10	3	33	—
		10	4	36	—
		10	5	23	2
	(*dia.* × *thickness*)				
Delamination	25 × 0.635	10	2	35	5
		10	3	37	6
		10	4	44	5
		10	5	46	6
		20	2	45	4
		20	4	57	4
		20	6	74	4
		20	8	57	4
Delamination	12.7 × 0.635	10	2	24	2
		10	3	50	2
		10	4	44	2
		10	5	34	2
		20	2	30	2
		20	4	64	2
		20	6	60	2
		20	8	68	2

TABLE 1—contd.

Defect	Dimensions (mm) (dia. × thickness)	Panel thickness (mm)	Defect depth (mm)	Corona (kV)	Microwave (mV)
Delamination	6.4 × 0.635	10	2	30	3
		10	3	55	2
		10	4	53	2
		10	5	55	—
		20	2	64	2
		20	4	69	2
		20	6	74	N.D.
		20	8	66	2

Notes
Corona voltage was obtained during raising of the voltage.
Microwave voltage was that at the detector after transmission—reflection—transmission through the sample. The signal was independent of defect depth, but dependent on defect size. Microwaves will be covered in Volume 2 of *Non-destructive Testing of Fibre-reinforced Plastics Composites*.

factor in the occurrence of discharges, and problems arose through the balanced build-up of charge on the test sample. Inception voltages (increasing potential difference) had a deviation of ± 1.2 kV for repositioned electrodes in similar positions, and ± 2.9 kV for repositioning anywhere on individual rings. The deviation in extinction voltage (reducing potential difference) was twice that for the inception voltage, but charge build-up was a factor here. The breakdown voltage for the rings was found to be in the range 47–52 kV, which imposed a lower limit of 0.2 volume percentage of voids (VPV) to avoid sample destruction. The examination of larger components (e.g. one-inch thick glass composite motor cases) would require corona test systems of up to 400 kV. The structure would need to be immersed in a non-reactive liquid medium of high dielectric strength, or the test electrodes would need to be encased in a high dielectric medium. They concluded that void size was related to the pulse amplitude, and that the number of voids was related to the pulse rate.

Rufolo and Winans [4] reported an early study to determine the resistance of plastic laminates to corona attack after periods of exposure, as indicated by changes in a variety of properties:

power factor, 1 kHz;
dielectric constant, 1 kHz;

volume resistivity;
surface resistivity;
step-by-step dielectric breakdown;
tensile strength;
flexural strength; and
flexural modulus.

A potential of 8 kV at 60 Hz, 25 °C and 50% relative humidity were selected, with additional tests of three materials at 96% RH. The variability of individual measurements precluded the detection of significant indications of material deterioration. Flexural strength and dielectric strength were most susceptible to degradation after corona exposure, but this selectivity was not explained.

Starr [5] carried out a comparison of the corona properties of 27 materials under various conditions. The materials which contained glass fibres (Teflon–glass laminates or melamine–glass laminates) were found to have consistent low corona-starting voltages (CSV) after water immersion. In particular, the melamine–glass was found to increase in dielectric constant after water immersion, which could explain the low value of CSV.

Lindsay and Works [6] reported that, after a void has been located by the corona test method, some indication of the size and shape of the void can be obtained by studying the characteristics of the corona pulse. By careful choice of the electrode system, a corona onset voltage was produced which was almost a linear function of the void thickness. The charge per pulse increased with increasing void thickness. The use of the Dakin–Malinaric capacitance bridge allowed differentiation between the corona produced on the surface and the corona resulting from voids within the specimen.

Magnaflux [7, 8] introduced a comprehensive non-destructive testing facility during the summer of 1963 at Aerojet-General Corporation in Sacramento to evaluate filament-wound Polaris A3 rocket motor cases. The CEBM (Corona, Eddy-current, Beta-ray, Microwave) system consisted of six main components used to automatically check the structural integrity, glass-to-resin ratios and inner wall thickness. The techniques were used as follows:

Corona: detection of gas bubble or voids, using a nominal 30 kV RMS voltage.

Eddy-current: measurement of rubber insulation thickness inside the case.

Beta-ray (strontium-90): detection of resin-rich or resin-starved areas, and hence volume fraction.

Microwave (12–18 GHz): voids, delamination or bubbles, by measuring the effective dielectric constant.

The system consists of a 10.7 × 1.83 m carriage capable of translation at 107–246 mm/min combined with rotation at 1–4 rpm.

There have been no significant reports of the use of the corona discharge NDT method for the last twenty years.

REFERENCES

1. J. A. Hendron, K. K. Groble, R. W. Gruetzmacher, G. O. McClurg and M. W. Retsky, Some nondestructive tests for filament-wound structures, ASTM STP 327, *Standards for filament wound reinforced plastics*, Silver Springs, June 1962, 261–272.
2. J. A. Hendron, K. K. Groble, R. W. Gruetzmacher, G. O. McClurg and M. W. Retsky, Corona and microwave methods for the detection of voids in glass–epoxy structures, *Materials Evaluation*, July 1964, **22**(7), 311–314.
3. B. E. Walker, C. T. Ewing and R. R. Miller, Nondestructive testing for void content in glass-filament wound composites, NRL Report 6775, 4 October 1968; NTIS AD 679 573.
4. A. Rufolo and R. R. Winans, Effects of corona on plastic laminates, *ASTM Bulletin*, May 1956, (214), 53–56.
5. W. T. Starr, Corona properties of insulating materials, *Electrical Manufacturing*, June 1956, **57**(6), 132–139.
6. E. W. Lindsay and C. N. Works, Corona discharge techniques as a nondestructive method for locating voids in filament-wound structures, ASTM STP 327, *Standards for filament wound reinforced plastics*, Silver Springs, June 1962, 273–286.
7. System inspects glass–epoxy rocket chambers, *Chemical and Engineering News*, 5 August 1963, 125–126.
8. CEBM unit to check Polaris cases, *Missiles and Rockets*, 12 August 1963, 39–40.

Chapter 7

Chemical Spectroscopy

JOHN SUMMERSCALES
Royal Naval Engineering College, Manadon, Plymouth, UK

and

DAVID SHORT
Department of Mechanical Engineering, Plymouth Polytechnic, Plymouth, UK

7.1. INTRODUCTION

Advances in the techniques of chemical spectroscopy combined with the modern powerful computing capability of contemporary electronics have extended the capability of these methods for use as non-destructive methods for the evaluation of fibre-reinforced plastics. This chapter surveys the use of several techniques for the monitoring of composites or their components. The subsections are as follows:

7.2. Nuclear magnetic resonance (NMR)
7.3. Electron spin resonance (ESR)
7.4. Raman spectroscopy
7.5. Fourier-transform infrared spectroscopy (FTIR)
7.6. Ultraviolet spectroscopy (UV)
7.7. Luminescence
7.8. Fracto-emission (FE)

7.2. NUCLEAR MAGNETIC RESONANCE

7.2.1. NMR background

Nuclear magnetic resonance in a complex molecule is generally achieved by placing the material in a strong constant magnetic field, and then applying a perpendicular radio-frequency (RF) alternating magnetic field. The nuclei absorb or emit energy at characteristic frequencies of the RF field, and the amount and frequency of this energy can indicate the chemical structure of the material under test [1].

The Bohr concept of the atom comprised negatively charged particles in orbits around a heavier positively charged nucleus. On this classical model, a particle of charge e and mass m, describing a circular orbit of radius r with a velocity of v, will possess an angular momentum p and a magnetic moment μ which are related [2] by the equations:

$$p = mvr \tag{1}$$

$$\mu = \tfrac{1}{2}evr \tag{2}$$

The gyromagnetic ratio of these two parameters, γ, is independent of the orbital details; thus

$$\gamma = \frac{\mu}{p} = \frac{e}{2m} \tag{3}$$

and in the presence of a magnetic field B the orbital particle is subjected to a torque proportional to μ and B. This results in a precession of the axis of rotation about the direction of the magnetic field with a frequency inversely proportional to p, such that

$$\omega_0 = \frac{\mu B}{p} = \gamma B \tag{4}$$

where ω_0 is the characteristic frequency, known as the Larmor frequency.

According to classical theory, the precessing particle should continuously lose energy, so that the orbit spirals inwards until μ and B become parallel. However, quantum theory indicates that only certain orbits are permitted and that transfer between these orbits is accompanied by a quantum change of energy. Any particle with spin should behave in a similar manner to the orbiting particle and have a characteristic Larmor frequency. If a large number of nuclei are forced to change simultaneously from one spin energy state to another, then a measurable induced magnetic field may be detected. By steadily varying either the magnitude of the magnetic field B or the frequency of the radio frequency field ω, the condition at which the nuclei change state may be detected as a weak RF signal or as a minute loss of power in the driver circuit. These two methods of detection are known respectively as nuclear induction or nuclear resonance absorption. The general term, nuclear magnetic resonance, abbreviated to NMR, is applied as a name for both manifestations.

In physical systems changes of state do not occur simultaneously, and decay or relaxation times are influenced by the environment of the nucleus. In NMR, two major groups of decay times are distinguishable:

T_1: the spin–lattice interaction, or longitudinal relaxation time; and

T_2: the spin–spin interaction, or transverse relaxation time.

The frequency of precession is an atomic constant characteristic of a particular nucleus, but the relaxation times are determined by the experimental conditions. Factors affecting the relaxation times include the uniformity of the applied field, shielding due to the bulk of the material, the time to achieve thermal equilibrium in the specimen and the boundary conditions arising from the sample container.

If the nuclear spins in a sample are aligned with a uniform magnetic field B_0 and a stronger magnetic field normal to B_0 is applied (B_{90}), then the nuclear moment will deflect; this is identified by the spin-lattice relaxation time T_1. The removal of the transverse magnetic field allows the nuclei to realign with the original uniform field B_0, identified with the spin–spin relaxation time T_2. In the realignment with the original field, the nucleus will perform the characteristic Larmor precession.

Recent advances in NMR spectroscopic techniques have led to the ability to measure chemical shifts in solids [3]. Combined with FT (Fourier transform) electronics, this has led to a massive expansion of the capabilities of NMR. The major new techniques include [4]:

(a) *Cross polarisation (CP) or proton-enhanced nuclear induction spectroscopy*

The chemical shift in solids is usually completely obscured by a larger homonuclear dipole–dipole coupling, resulting in broad Gaussian lineshapes. The homonuclear coupling can be removed by isotopic dilution of the observed spins, but there is a significant loss of sensitivity as a result of this approach. In proton-enhanced nuclear induction spectroscopy the observation of dilute spins S, in a sample with abundant spins (usually protons) I, has made measurement of the chemical shift tensors of the S-spins available in solids. Dilution occurs naturally for ^{13}C nuclei as they are only ~1% abundant in nature. Sensitivity is enhanced by *cross-polarisation* (CP) from the *I*-system to the *S*-system through energy-conserving transitions induced by the heteronuclear dipole–dipole coupling. The polarisation of the *S*-spins can be enhanced by γ^I/γ^S, which is 4 for $^{13}C-^1H$ and is 6.5 for $^2H-^1H$.

(b) *Magic-angle spinning (MAS)*
The broadening of NMR spectra in solids can be reduced by rotating the sample at the magic-angle β_m between the principal axis of the second-rank tensor describing the secular terms of the Hamiltonian and the *z*-direction of the external magnetic field. It was originally believed that the sample would have to be rotated at a frequency higher than the magnitude of the broadening, which is impossible in the general case. It was subsequently discovered that rotation of the sample at a frequency much lower than the inhomogeneous linewidth has a dramatic effect on the inhomogeneously broadened linewidth. The frequency of rotation must only exceed the homogeneous linewidth of the spin packets which comprise the inhomogeneous resonance, and sharp sideband spectra are obtained with 'slow' spinning. When the homonuclear dipole–dipole coupling of the observed *S*-spins is small, the inhomogeneous broadening due to chemical shift can be easily removed. This is the natural case for ^{13}C and ^{31}P in solids. Cross-polarised magic-angle spinning (CP-MAS) ^{13}C-NMR thus rapidly became a standard tool for the chemist.

(c) *Multiple pulse line narrowing*
The application of many RF pulses combined with suitable time delays allows the removal of homogeneous dipolar broadening of the NMR spectra of solids. This technique has further been combined with magic-angle spinning to obtain liquid-like spectra, even for protons where the broadening is 50–100 kHz. Pulse sequences eliminate higher-order correction terms and reduce the effects of weak and imperfect RF pulses.

(d) *Double quantum Fourier transform NMR*
The 'forbidden' ($\Delta m = 2$) double quantum transitions in the three-level system of a nucleus with $I = 1$ can be induced with weak RF irradiation ($\omega_1 > \omega_0$). Observation of deuterium double quantum transitions was developed for pulsed FT-NMR with heteronuclear double resonance decoupling, allowing the much smaller chemical shift tensor in solids to be measured. This allows a novel approach to high resolution NMR in solids.

(e) *Deuterium double quantum decoupling*

High resolution proton NMR in solids can be approached by the observation of spectra from isotopically dilute protons S in a deuterated solid. Dilution of the protons reduces the homonuclear dipole coupling to a very small value. Radio-frequency decoupling is then used to remove the broadening of the proton spectra due to the heteronuclear dipole coupling with the deuterons. The homonuclear dipole coupling of the deuterons is small and can be neglected.

Hackelöer *et al.* [5] have recently described a combined servohydraulic tension/compression machine and pulsed NMR spectrometer which enables nuclear spin relaxation rates to be recorded simultaneously with stress–strain data. The system has been used to investigate dislocation motion in both non-metallic and metallic samples, but no work on fibre-reinforced plastics has yet been reported.

7.2.2. NMR spectroscopy of composites and reinforcement fibres

As early as 1963, Epstein and Weinberg [6] conducted an experimental investigation of both NMR and NQR (nuclear quadrupole resonance, see Section 7.2.4) spectroscopy of fibreglass. The NMR method was found to have potential for a number of NDT purposes. During the cure of an epoxy resin, the proton resonance showed extensive broadening (over 200% increase). The resonance frequency absorption bandwidth is inversely proportional to the cube of the nearest neighbour distance. During cure of the resin, the molecules draw significantly closer together with a consequent reduction of the nearest neighbour distance, accompanied by the restriction of molecular motion.

NMR spectra were also recorded for the ^{29}Si nuclei in the E-glass fibre roving. Despite the presence of silicon in the probe, the ^{29}Si from the glass fibres was apparent as a shoulder superimposed upon the response of the empty probe (Fig. 1). It was thus concluded that excellent sensitivity could be achieved when using a silicon-free probe. The technique could be applied to location and identification of glass fibres and for quantitative determination of fibre integrity and volume fraction.

Stepanov *et al.* [7] used a 12 MHz spectrometer to examine 6–7 μm diameter aluminoborosilicate glass fibres. During the fibre forming process there was an abrupt drop in temperature. The structure of the material was 'frozen' to restrain the restructuring of BO_3 groups into BO_4 tetrahedra as would otherwise occur during slow bulk annealing. Three fibres were considered: the original, those annealed at 300 °C and those annealed at 500 °C. At the higher temperature, the structure was allowed to ap-

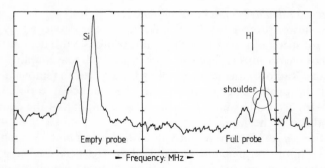

FIG. 1. Nuclear magnetic resonance response of glass fibres, observation of ^{29}Si (from Epstein and Weinberg [6], copyright Ford Motor Company Aeronutronic Division, 1963).

proximate to the equilibrium state characteristic of the source glass as seen from the form of the NMR spectra. The BO_3 groups were capable of forming chain structures possessing greater strength, viz: 2.48 GPa (original), 2.44 GPa (300 °C anneal), 1.55 GPa (500 °C anneal).

Stepanov and Novikov [8] used ^{11}B-NMR, recorded as above, to assess the ratio of boron atoms in fourfold or in threefold coordination. The ratio of the BO_4 groups to BO_3 groups was expressed as N_4/N_3, and was found by double numerical integration of the derivative absorption curve. The relative variation of N_4/N_3 between specimens was found directly from the experimental spectra by the ratio of the amplitudes of the narrow central resonance (N_4) and the broad distorted resonance (N_3). Spectra were recorded with a fixed generator frequency of 14 MHz. Upon heating the glass fibres there was a regular increase in fourfold coordination relative to threefold coordination. Similar treatment of bulk glass of the same composition did not lead to changes in the ^{11}B environment. Brief etching of the glass fibre surfaces with 0.2% HF reduced the ratio N_4/N_3 for heat-treated fibre. In air, at between 300° and 600 °C, there was a regular decrease in the amount of oxygen in the heating vessel. When the glass fibres were heat-treated in argon, the coordination number of the ^{11}B atom did change, although less markedly than in air. It was stated that there was 0.4% oxygen in the argon. The changes in the coordination number of boron are concluded to be due to surface processes in which oxygen from the air interacts with coordinately unsaturated centres.

Guseva et al. [9] used ^{11}B-NMR spectra to examine the changes to Al–B–SiO_4 glass fibres during heat treatment between 200 and 500 °C. The donor/acceptor interaction between atmospheric oxygen and the surface of the glass fibres causes a change in the coordination number of the boron to occur. The conversion of BO_3 groups to BO_4 groups was followed by this technique and correlated with the amount of oxygen absorbed by the fibre. The reduction in fibre strength and the increase in the volume of defects (as measured by adsorption and porometry) are associated with the interaction of the fibre surface with atmospheric oxygen. Experiments in argon and *in vacuo* showed a far lower increase in the volume of defects. Even the small amounts of air in 99% pure argon, and an insufficiently high vacuum, were shown to lead to a visible reduction in strength during the heat treatment of the fibre above 200 °C. This mechanism of interaction between coordinately unsaturated centres and the atmosphere probably takes place in other silicate fibres during heat treatment.

Stefan and Williams [10, 11] examined the spin–lattice relaxation time (T_1) and the spin–lattice relaxation time in a rotating frame ($T_{1\rho}$) for a variety of materials, including glass fibres, in bisphenol-A polycarbonate. It was found that multiple transitions in the composites were not always combinations of the transitions in the individual components. A broadening of the T_1 minimum associated with reorientation of the methyl group indicated a wider distribution of relaxation times for this transition and thus greater non-equivalence of the side groups. The nuclear spin relaxation in the rotating frame showed two different relaxation rates, which were interpreted as due to two phases in the polymer: one near the filler, which imposes motional restraints on the polymer chains, and another with motions equivalent to those in pure bulk polymer.

Fabulyak *et al.* [12] investigated the dielectric relaxation and NMR spin–lattice relaxation of acrylate-epoxy-styrene compounds in bulk and in the surface layers of glass fabric. The glass fibres were degreased with toluene, heated to 400 °C, vacuum dried at 110 °C, fully cooled and impregnated with the cold-setting resin, then cured in sealed ampoules. The results of the studies of molecular mobility in bulk by dielectric measurement were corroborated by the spin–lattice times. The downward shift in chain relaxation temperatures due to impurities on the glass surface was observed both in dielectric and NMR measurements. Heat treatment of the compounds was shown to reduce the shift from 21° for an unheated surface to 12° in a treated surface layer. It was concluded that the constraint of molecular mobility was greater in flexible polymers than in rigid macromolecular chains.

Lipatov *et al.* [13–15] have studied the changes in molecular mobility in a polymer filler–matrix system during solidification. At low degrees of solidification the fabric acted as a filler and the mobility of the matrix was reduced. At high solidification the molecular motion of the polymer fabric was reduced below that of the resin. The implication was that the mutual interaction of the components affected the molecular mobility of both components as indicated by dielectric relaxation and NMR spin–lattice relaxation times. Lipatov postulated that the nature of the interface was independent of the chemical nature of the surface unless there was a specific interaction between the filler and the matrix. An increase in T_g of the filled polymer suggested that the motion of the backbone chain was restricted by the solid surface, with a consequent reduction of the number of possible configurations. Consequently a layer of restricted loosely-packed polymer chains forms near the filler surface. It was suggested that side chain motion near the filler was more flexible, while the main chain motions were still restricted. This was supported by NMR and dielectric measurements [14] on the polyamide fibre-reinforced epoxy, and by the existence of a rarified interphase as supported by density measurements [16] of filled polymethylmethacrylate.

Kaelble and Dynes [17] have reported that Witt has shown that the presence of moisture affected both the free induction decay (FID) and the T_1 and T_2 relaxation times in graphite/epoxy composites. The NMR technique indicated that the absorbed water was tightly bound to the epoxy resin, which is in agreement with the high temperature ($> 120\,°C$) required for internal moisture to affect the molecular motion and low frequency damping of the composite.

King et al. [18] used a 30 MHz transient hydrogen NMR system to study a single graphite/epoxy specimen. The dried specimen had a single free induction decay time of 8 μs. After exposure to 8% RH at 52 °C, the 8 μs part of the FID did not change, but a component with a longer relaxation time occurred which increased with the duration of the exposure (10 μs after 24 h, 24 μs after 72 h and division into two relaxation times after 336 h). A weight loss method indicated 0.8% moisture content in the composite after 336 h.

Matzkanin [19, 20] conducted an investigation of the feasibility of using 30 MHz pulsed hydrogen NMR to nondestructively determine the amount of moisture in organic matrix composites and the extent of moisture-induced mechanical degradation. Two composite material systems were studied: 8-ply $\pm 45°$ S2-glass fibre in SP-250 resin and Kevlar 49 fibre in Reliabond 9350 resin. Long-term environmental conditioning for four months at 95% RH and 52 °C was undertaken. Reductions in tensile strength were 4.3% in a fibreglass with 1.3% moisture content and 14%

in a Kevlar composite with 4.6% moisture content. Absorbed moisture levels as low as 0.2% produced measurable NMR signals, consisting of distinct multiple components attributed to moisture in various states of molecular binding. Good correlation was obtained between the NMR signal output and the dry-weight moisture percentage. Experiments on fractured specimens, subjected to additional environmental exposure, showed that NMR is capable of distinguishing the free moisture entering a composite through cracks and fissures from that moisture absorbed into the composite structure. Analysis of the NMR signals indicated that plasticising of the resin matrix, rather than microcracking, was the most likely cause of mechanical degradation. Experiments on the fibres showed that no proton-NMR signal was observed from the glass fibre, although moisture absorbed by the Kevlar fibre could be detected.

The important feature of the Matzkanin NMR results is that not only can the amount of moisture be determined, but the method can also distinguish between water in different binding states. Pulsed NMR free induction decay (FID) signals were analysed following a 90° RF pulse. This measurement allows a determination of the nuclear spin–spin relaxation time (T_2) which provides information on molecular coupling and on nuclei in different chemical states. Signal average techniques were used to reduce the signal-to-noise ratio. The NMR FID signals were generally characterised by a large-amplitude fast-decaying component ('solid component': $T_{21} = 6$ μs) identified with chemically bound structural hydrogen in the composite, and a slower-decay lower-amplitude component associated with the absorbed moisture. The Kevlar composite (1.36% moisture) exhibited three slower relaxation times ($T_{22} = 80$ μs, $T_{23} = 157$ μs and $T_{24} = 253$ μs) whilst fibreglass (0.37% moisture) showed only two relaxation times ($T_{22} = 28$ μs and $T_{23} = 71$ μs). The absence of hydrogen from the glass fibre structure is associated with the reduced number of relaxation times. As the moisture levels increase, the time constants also increase, indicating that motion of the moisture is increasingly constrained, as shown in Table 1.

Matzkanin [21] has reported results on glass fibre and Kevlar fibre-reinforced plastics, and lists the problems which NMR is capable of addressing as including the measurement of moisture, environmental degradation, modulus variations, degree of cure, impact damage and hydrogen concentration variations.

Lawing et al. [22] used a broadline NMR spectrometer to investigate the temperature dependence of water-soaked graphite/epoxy (TGDDM/ DDS) cylindrical unidirectional composites. The narrow line consistently

TABLE 1
Variation of Relaxation Times with Moisture Content

Kevlar composite					Parameter	Fibreglass composite				
0.49	1.36	2.00	2.83	4.60	% moisture	0.19	0.37	0.62	0.78	1.27
70	75	82	113	178	$T_{22}(\mu s)$	13	33	35	45	54
175	157	175	200	230	$T_{23}(\mu s)$	62	70	76	99	125
—	210	228	235	245	$T_{24}(\mu s)$					

observed in the broadline NMR spectra in the temperature range (25°, −5°, −20°, −42 °C) studied arose from an intermediate state between free liquid water and bound solid-state water, probably due to hydrogen bonding with the polymer. As the temperature was lowered, some of the intermediate-state signal became indistinguishable from the broad background line, due to the protons in the resin. The intermediate signal reappeared on reraising the temperature, and the water line was still present below 0 °C. Ratios of the areas beneath the narrow and broadline NMR absorption were a function of temperature:

Temperature (°C)	25	−5	−20	−42
Area ratio (%)	8.5	3.3	2.5	2.3

The room-temperature ratio of 8.5% corresponds to a ratio of mobile water in the composite of 1.2% by weight. The spectrum was isotropic upon rotation of the fibre axis with respect to the direction of the externally applied magnetic field, which is in marked contrast to studies of certain other fibre systems reported elsewhere.

Fuller et al. [23–25] continued the above study and examined four environmental cases: soaked in H_2O, soaked in D_2O, dried, and not treated. Proton spectra were obtained at 56.4 MHz fixed radio frequency, with signal/noise enhancement by time averaging. The first group of samples (cured resin only) at room temperature were examined dry and soaked. The dry samples had a single broad peak due to protons in the solid state. The peak width was 10 G. The soaked samples exhibited a broad peak and a narrow line due to the water absorbed by the sample. The narrow line was 1 G wide, which indicated that the water was in an intermediate solid/liquid state. At elevated temperature (68 °C) a narrow line became visible for the dry samples, and the water line of the soaked samples was sharpened by the increased mobility of the water.

In Fuller's second group of samples, graphite–epoxy composites, the spectra were measured as the angle between the fibre and the magnetic field was changed. There was no change in the spectra of either dry or soaked samples with angle. It was concluded that either water was only associated with the bulk resin, or water associated with the fibres did not have a preferred orientation. In a third group of experiments, a sample was dried and soaked in D_2O, whereupon a narrow line appeared in the proton spectrum. It was concluded from this that hydrogen exchange was taking place between the resin and the absorbed water, probably at the hydroxyl group of the opened epoxy ring. The frequency of the band in the infrared spectrum (wave number 3300) is consistent with a hydroxyl stretching frequency. It may of course be that the exchange occurs in unreacted DDS curing agent.

Batra and Graham [26] have shown that continuous wave (CW) NMR can be used to measure small traces (a few mg) of moisture in composites of both graphite–epoxy and glass–epoxy. A radio-frequency magnetic field (H_{rf}) of 15.9932 MHz, simultaneous with a static field (H_0) of 3.7558 kG, was applied with a parallel sweep field of ± 50 G. H_{rf} was orthogonal to H_0. In composites containing moisture, the protons are loosely bound in the water when compared to the protons in the epoxy matrix. Tightly bound protons have strong coupling of their magnetic moments. Consequently their interactions are strong and the line width is fairly large. For protons in water, the line widths are very small due to the rapid tumbling of the nuclei which largely averages out the dipolar fields generally responsible for broadening the line. The NMR spectra of composites containing moisture consists of a very sharp proton line superimposed upon a very broad background signal. Since the number of nuclei is proportional to the area under the line in the NMR absorption curve, the moisture content in a given sample can be estimated quantitatively from the NMR spectra. A plot of resonance intensity against milligrams of water showed a linear relationship in the range 0–40 mg.

Chiang et al. [27] used high resolution solid-state cross-polarisation (CP) magic-angle spinning (MAS) ^{13}C-NMR spectroscopy (operating frequency 37.7 MHz, proton decoupling power 20 G, 2.1 kHz spinning) for the study of organofunctional silane coupling agents, both in bulk and adsorbed on glass powder surfaces. The resonance peaks of low-cure and high-cure polyaminopropylsiloxanes were chemically shifted from one to the other. The hydrogen-bonded propyl chain of APS (γ-amino propyltriethoxysilane) produced an upfield shoulder on the central methylene carbon resonance peak in the CP-MAS experiment. Three molecular

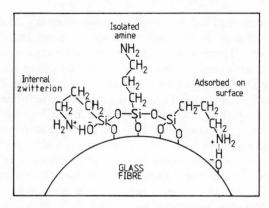

FIG. 2. The three molecular structure models of the APS coupling agent on glass fibres (from Chiang, Liu and Koenig [27], reproduced from *Journal of Colloid and Interface Science* with the permission of Academic Press).

structure models of the APS on glass were proposed, involving a hydrogen bonding interaction between the amine and the SiOH groups (Fig. 2). The chemical shifts induced by the electric field effects on the β-carbon nucleus for different APS isomers were calculated and found to agree with the observed resonances. The technique was capable of resolving less than a monolayer equivalent of coupling agent on the glass surface.

Cholli et al. [28] used the same CP-MAS ^{13}C-NMR technique for the characterisation of different regions in fibre-reinforced plastics, especially:

(a) studies of coupling agents on silica surfaces;
(b) curing of epoxy at elevated temperature; and
(c) the $T_{1\rho}$ relaxation time of the matrix in the presence of fibre.

In the first group of studies, an upfield shift of as much as 2 ppm was observed for all the propyl carbons of the silanes when attached to the surface, relative to the position in bulk polymer, probably due to the increased steric hindrance at the binding site. Surface impurities, inhomogeneities and hydrogen bonding all potentially contribute to the line broadening. In the second group, the decrease of the resonance intensity for the epoxy ring carbon (50.2 ppm) and the increase in the oxymethylene carbon resonance (71.3 ppm) were plotted during cure. The relationship between the two parameters became linear above 57% extent of reaction. The non-linearity in the early stages of the cure could be due to either the reaction mode or the efficiency of cross-polarisation, which may increase as the reaction proceeds. In the third group, the ^{13}C $T_{1\rho}$ (spin–lattice relaxation time in a rotating frame) of the matrix appeared to be reduced in the presence of the fibre. This would be indicative of the imposition of high constraint of molecules on or near to the fibre surface.

Evdokimov et al. [29] studied polyheteroarylene (PHA) fibres of 11–13 μm diameter in ÉL-20 or DÉG-1 epoxide resins cured with triethanolamine titanate at 165 °C for 3 h. The NMR spectra were recorded with an operating frequency of 29 MHz, a modulation frequency of around 20 Hz and a sweep rate less than 4 G/min on a modernised RYa-2301 spectrometer with an autodyne detector. In the spectra of PHA and of pure binder, a narrow line was not observed in the range of temperature studied. At a temperature of 100 °C the width of the fibre signal had changed little, but the width of the signal for pure binder at 100 °C corresponds to the width of a narrow line observed in the composite at 60 °C. As the composite temperature further increases, the NMR line narrows and increases in intensity. This narrow component of the spectrum was ascribed to the increased molecular mobility in 'defect regions' of the binder existing at the fibre–binder interface. The disruption of packing in the binder macromolecules by the introduction of the fibres caused a 20 °C drop in the glass transition temperature.

Sugiya et al [30] studied Kevlar and Conex fibres using a broadline NMR apparatus with a Robinson-type detector operated at 17 MHz. The radio-frequency magnetic field was kept as low as possible to avoid saturation, with peak-to-peak modulation amplitude of 2 gauss. The alignment angle (ω) was taken to be zero when the main magnetic field was parallel to the fibre axis and 90° when the field was perpendicular. The fibres were:

Kevlar (DuPont via Toray Co)— Continuous filament yarns of PPPT: poly(para-phenylene terephthalamide). Equivalent to Monsanto PaBH-T × −500.

Conex (Teijin Co) —Single spun yarns of PMPI: poly(meta-phenylene isophthalamide). Equivalent to DuPont Nomex.

The fibres were examined at several temperatures by measuring the anisotropy of the second moment of the proton NMR absorption line. Although the motional narrowing was very slight for Conex in the temperature range $-150°$ to $+170\,°C$, it was appreciable for Kevlar between $-50°$ and $+150\,°C$. It was suggested that the narrowing was due to a thermally stimulated 'flip-flop' motion of the phenylene rings about their para-axes, with a frequency of 100 kHz. Consideration of the barrier heights for rotation of the phenylene rings on both sides of the amide group suggest that the flip-flop motion will be realised, more easily than anticipated, in one half of the phenylene rings. Flip-flop motion in the isophthalamide ring is impeded by the asymmetry of the ring relative to the axis.

Hempel and Schneider [31] examined a highly oriented Terlon polyamide fibre using ^{13}C-NMR with a Larmor frequency of 22.64 MHz and a radio-frequency magnetic field of 4 mT, with high power proton-decoupling. With the fibre axis parallel to the magnetic field, the ring carbons adjacent to the carboxylic groups caused a line at the position

$$\sigma_s = \sigma_{11} \cos^2 \psi + \sigma_{33} \sin^2 \psi \tag{5}$$

where signals in this spectrum were restricted to the range σ_{11} to σ_{22}, leading to a value of ψ within the range $0°$–$35°$. With the fibre axis perpendicular to the magnetic field a doublet was revealed, with peaks at

$$\sigma_{D1} = \sigma_{22} \tag{6}$$

$$\sigma_{D2} = \sigma_{11} \sin^2 \psi + \sigma_{33} \cos^2 \psi \tag{7}$$

The absence of any peak between σ_{22} and σ_{33} restricts ψ to values between $55°$ and $90°$, or $0°$ if $\sigma_{D2} = \sigma_{33}$. Because of the contradiction between the two cases above the only possible value of ψ will be zero. The implication of this is that the angle between the fibre draw direction and the plane of the phenyl rings is zero. It was suggested that the polymer conformation was a mixture of cis- and trans- forms of the stretched chain, with some distortion of the bond angle of the ring-carbon adjacent to the carbonyl atom.

Goryainov and Koltsov [32] used wide-line NMR to study the spectra obtained for different angles between the magnetic field and the axis of partially crystalline polypyromellitimide fibres in the temperature range 77 to 623 K. At 77 K the second moment of the spectrum was independent of temperature and determined by the mutual disposition of the fixed protons. At 93 K there was a small decrease in both linewidth and second moment accounted for by torsional vibrations of the rings with a frequency of $\geqslant 10$ kHz. This did not affect the anisotropic intramolecular component of second moment, but was due to the intermolecular component. From 423 K until 473 K the decrease reduced, then terminated in a plateau. In an amorphous polyimide containing a complex dianhydride part this reduction began at 243 K. In the high temperature region, a further decrease of the second moment was observed for two polyimides, which may be attributed to the occurrence of segmental or rotational mobility due to the transition from the glassy to a highly elastic state. For anisotropic polyimides, the second moment curves were similar in all cases at liquid nitrogen temperature, room temperature and 573 K, indicating that the main form of molecular motion, over a wide range of temperatures, was the torsional vibration of the phenyl rings. It was anticipated that steric hindrance would limit the vibration to between 30° and 40°. For weakly oriented amorphous polyimides the second moment had the same value at all angles within the limits of instrumental error. Assignment of mechanisms is more difficult in this case because of the possibility of more complex forms of molecular motion.

Alston [33] studied the thermal degradation of Celion 6000 carbon fibres in PMR-15 polyimide resin using an 80 MHz ^{13}C FT-NMR spectrometer. The composites were cured at 589 K and exposed to air at the same temperature for prolonged thermooxidative ageing. Overall resin weight loss was determined by chemical separation of the matrix to the monometric components. The isothermal weight loss of the resin was correlated to changes observed in the spectra obtained by FT-IR and FT-NMR. The observed weight loss of 4,4′ methylene diamine was linear in respect of the square root of the exposure time, suggesting a diffusion-controlled thermooxidative reaction. Intermediate oxidation products were not detected in the composite. After 1200 h the nadic anhydride cross-link component began to degrade with an associated weight loss which rapidly increased to become a significant contributor to the loss.

Lauver et al. [34] used NMR spectroscopy to assess chemical changes to polyimide resin monomer solution during ambient storage and at 5 °C and −18 °C. Measurements were obtained on a commercial 60 MHz CW-NMR instrument with samples dissolved in deutero-chloroform and referenced to tetramethylsilane. The shelf-life of the monomer solutions was significantly increased at reduced storage temperatures. Unidirectional 8-ply HTS-carbon fibre composites prepared from the aged solutions had properties (flexural strength and interlaminar shear strength at ambient and 316 °C) typical of high quality panels. The only variations noted in processing of the aged solutions were slight changes in the resin flow and in fibre wetting.

Kowalska and Wirsén [35] analysed commercial epoxy prepregs from eight different manufacturers using NMR, mass spectrometry and gel permeation chromatography. Resin components could be analysed qualitatively with high sensitivity even for trace amounts. Quantitative proportions of the main components were rapidly estimated by NMR regardless of the degree of precure. Proton NMR was undertaken with a 60 MHz continuous wave spectrometer. ^{13}C-NMR used a 20 MHz spectrometer, which was proton-decoupled. Both were internally referenced to TMS (tetramethylsilane). Proton NMR of some prepregs dissolved in deuterated chloroform gave a good picture of the main components present, but was unsatisfactory for quantitative measurement because of poor solution of DDS and of broadening of peaks due to precure, in all solvents. Carbon NMR was used qualitatively with deuterated chloroform solution and indicated that six of the prepregs were based on TGMDA/DDS. Quantitative ^{13}C-NMR was undertaken using 5–10% acetone solutions extracted for 24 h. A molar relation of TGMDA/DDS was obtained by comparison of the 130.1/129.5 ppm peaks respectively, which was independent of the degree of precure and was linear in the range 0.1–3.0 mole TGMDA/1 mole DDS. An accuracy of $\pm 10\%$ could be obtained in 30 minutes by comparison of the peak heights, but to achieve an accuracy of $\pm 1\%$ required several hours and multiple scans, integration contra peak heights, pulse delay, decoupler mode and acquisition times to be taken into account.

Happe et al. [36, 37] used an FT-NMR spectrometer to obtain ^{1}H, ^{11}B and ^{19}F spectra at frequencies of 200.071, 64.190 and 188.228 MHz respectively for the characterisation of cure in commercial epoxy prepregs. The systems studied were Fiberite 934 and Hercules 3501 carbon fibres in TGDDM/DDS resin with boron trifluoride catalysts. The experiments revealed a significant amount of detail about the variability and composition of the curing agents under the influence of heat. The work was directed towards the clarification of the reactions and would be more appropriate for prefabrication QC testing of the prepregs.

In addition to the literature reviewed here for NMR spectroscopy of polymer composites there is a considerable literature on the use of this technique for the characterisation of cure of thermosetting polymers in the absence of fibrous reinforcements. A very limited selection of these papers are listed as references [38–52].

7.2.3. NMR imaging: computer tomography

The potential of nuclear magnetic resonance as a rapid, biologically inert, diagnostic imaging system for the study of in vivo biochemical processes is rapidly approaching commercial production. Damadian [53] introduced a system of focusing NMR (FONAR imaging) in which the variability of nuclear magnetic relaxation times of the hydrogen nuclei in different cellular environments is used. Lauterbur [54] simultaneously introduced 'zeugmatography', utilising the spatial mapping of the NMR response of hydrogen-containing fluids in the object. Mansfield [55] and Hinshaw [56] soon followed these methods with fast-scan imaging by a diffraction method, and sensitive point imaging, respectively. All these techniques have aroused commercial interest [57] for both whole-body imaging and local spectrum (TMR: topical magnetic resonance) systems. The use of NMR as a medical imaging technique is covered in greater detail in Refs 58–62.

Rothwell et al. [63] used NMR imaging to nondestructively determine the distribution of water in 77% unidirectional glass fibre-reinforced bisphenol–epoxy composites. Only three samples were examined:

(a) aromatic amine cure, small shrinkage cracks at rod centre, 70 days in 93 °C water bath;
(b) anhydride cure, 90 days in 93 °C water bath; and
(c) anhydride cure, not exposed to water, vacuum-dried at 100 °C, control sample.

The NMR image of composite (a) clearly showed a non-uniform distribution of absorbed water, preferentially positioned through the rod centre. In composite (b), the water was found to be non-uniformly distributed around the rod perimeter, with marginally lower concentration of absorbed water at < 1% by weight. The dried composite, (c), did include some residual moisture, but at a concentration very much less than 1%. Despite the limited number of tests, the results presented a clear demonstration of the heterogeneous absorption of water by the composites. The mechanism and nature of this process appear to depend strongly on the chemical and physical details of the material. In addition to the spatial mapping, it should be possible to distinguish interstitial water from tightly bound water through consideration of the different NMR relaxation times.

NMR instrumentation is developing rapidly and the scope for comprehensive monitoring of the condition of the composite will increase with the separation of the different relaxation times, multinuclear spectroscopies (^{1}H, ^{11}B, ^{13}C, ^{19}F, ^{31}P etc.) and advances in signal processing.

7.2.4. Nuclear quadrupole resonance

The nuclear quadrupole moment is connected with the spatial orientation of the nuclear spin and gives a measure of the lack of sphericity in the distribution of electric charge within the nucleus. In an inhomogeneous electric field with symmetry lower than cubic, such as exists in certain crystals and molecules, the nuclear quadrupole moment has energy states that correspond to various orientations of the nuclear spin with respect to the axes of symmetry of the local electric field. The splittings between these levels range from a fraction of a megahertz to a few hundred and, occasionally, a few thousand MHz, depending on the nucleus and the nature of the sample. When a magnetic field is applied to a sample containing nuclei with quadrupole moments, there is a continuum of situations between the pure Zeeman resonance (when the symmetry of the electric field is cubic or higher) and the so-called pure quadrupole resonance (when the applied magnetic field is vanishingly small). There is nothing special about the latter situation except the possibility of obtaining sharp spectral lines for that case with polycrystalline samples [64].

Epstein and Weinberg [6] used nuclear quadrupole resonance to monitor shrinkage stresses and internal stresses caused by an external applied force. Finely divided cuprous oxide was used as a tracer because

(a) its frequency–pressure coefficient of 2.5×10^{-3} kHz/psi is constant in the range $-77°$ to $+98\ °C$;
(b) it is well characterised and readily available; and
(c) it is reasonably inert and easily handled.

Four volume fractions of cuprous oxide (2.2, 6.3, 16.7 and 37.5%) in epoxy resin were prepared and cured simultaneously. The signal from the 2.2 v/o material was too weak for good frequency measurement with the equipment available at the time, 1963. In all cases the NQR frequencies for the resin samples were higher than those of pure tracer material, as expected, due to the pressure caused by the shrinkage stresses during cure. The shrinkage stress at the interface, determined by extrapolation to zero-percent tracer, was equivalent to 33 MPa. Glass fibres subjected to stresses of this magnitude could readily influence the performance of the composite. Externally applied clamping stresses were also detected as an NQR resonance-frequency shift.

Hewitt and Mazelsky [65, 66] examined the NQR of dilute inert fillers in several polymers and adhesives as a function of compression and tension. The material used as a probe was the ^{63}Cu nucleus in cuprous oxide (Cu_2O) powder of analytical grade, with 50 μm diameter crystallites. The resin system was Epon 828 cured with Versamide 140 catalyst. The change in the NQR response was proportional to the strain applied to the host material. The intensity of the NQR tracer signal was in substantial agreement with the reduction expected from the tracer dilution. Polymer rods reinforced with glass fibres and glass cloth were tested in tension, allowed to relax for 24 h, then tested in compression to the same magnitude (12.4 MPa). No systematic NQR response was observed at these low strains. Repeated stress cycling caused some mechanical degradation of the glass–resin bond, accompanied by a shift of the NQR frequency under stress and broadening of the resonance line. The results were not reproducible and depended on the state of degradation in the particular sample. The same general trends of line broadening with increasing stress and a possible increase in the resonant frequency were also observed in specimens manufactured with simulated flaws.

7.3. ELECTRON SPIN RESONANCE

7.3.1. Background [67]

Any atom or molecule which possesses one unpaired electron is termed a 'free radical'. The magnetic properties inherent in free radicals, because of the presence of these unpaired electrons, can be utilised through the techniques of electron spin resonance (ESR) spectroscopy. A free electron is a spinning charged body with an associated magnetic moment, μ, given by:

$$\mu = g\beta S \qquad (8)$$

where β is the Bohr magneton (9.274×10^{-24} J/T) and S is the spin vector of the electron. The g-factor, or spectroscopic splitting factor, for a free electron is 2.0023. An electron contained in a magnetic field H_0 acting in an arbitrary reference direction Z interacts with the field with an energy given by:

$$E = -\mu_z H_0 = g\beta M_s H_0 \qquad (9)$$

where M_s is the magnetic spin quantum number with a value of $\pm\frac{1}{2}$. This interaction gives rise to two energy states dependent on whether the magnetic moment vector is aligned parallel or antiparallel to the field. The separation of these two states in energy terms is given by:

$$\Delta E = g\beta H_0 = h\nu \qquad (10)$$

At microwave frequency ($\nu = 9.5$ GHz) the required magnetic field is easily obtained ($H_0 \approx 3300$ gauss). In an ESR experiment the sample is irradiated with microwave radiation at constant frequency, while the magnitude of the magnetic field (H) is varied. When the conditions of eqn (10) are satisfied, energy is absorbed which can be detected electronically.

Conversely the magnetic field may be held constant while the microwave frequency is varied.

ESR spectroscopy relies on the interaction of the magnetic moment of the unpaired electron with the magnetic moment of the nucleus of the radical. For example, in a hydrogen atom the unpaired electron will be in the spherical 1s orbital centred on the proton, and will have a spin $I = \pm\frac{1}{2}$. The magnetic moment vector of the proton will be aligned parallel ($m_I = +\frac{1}{2}$) or antiparallel ($m_I = -\frac{1}{2}$) to the applied magnetic field, H_0. If the magnitude of the field resulting from the magnetic moment of the proton is denoted H_1, then the unpaired electron will be subject to two fields: $H_0 - H_1$ and $H_0 + H_1$. These two values of applied field can thus satisfy eqn (10), and the single absorption line is thus split, such that the intensity and the area under each line reflects the concentration of electrons subjected to each field. The spacing between the lines is known as hyperfine splitting (HFS), commonly denoted a and expressed in gauss or MHz (the conversion factor is 1 gauss = 2.803 MHz).

Electron spin resonance was first observed in 1945 by Zavoisky. The technique is also known by the synonyms 'electron magnetic resonance' (EMR) and more commonly 'electron paramagnetic resonance' (EPR) [68].

7.3.2. ESR of carbon fibres

Robson et al. [69] made a systematic study of bundles of PAN–carbon fibre precursor using a conventional X-band ESR spectrometer with 100 Hz phase-sensitive detection. The g-factor of the resonance line was observed while the static magnetic field was rotated about an axis perpendicular to the fibre axis. The g-factor for a fibre treated to 2700 °C exhibited a $\cos^2\theta$ angular variation with a maximum ($g = 2.0076$) when the magnetic field was perpendicular to the fibre axis and a minimum ($g = 2.0020$) when these axes were parallel. The ESR results confirmed earlier x-ray and electron diffraction results which concluded that the crystallites within the fibre were preferentially aligned with their basal planes lying parallel to the fibre axis. The g-values were obtained for fibres heat-treated to temperatures between 1000 and 2700 °C. The anisotropy of the g-values increased with the heat-treatment temperature, echoing the growth and ordering of the crystallites within the fibre. This study was continued with extension of the temperature range to 2800 °C [70]. The anisotropy was found to appear at a heat-treatment temperature (HTT) of 1750 °C and to increase with temperature. The magnitude of the resonance line illustrated that crystallite development was, however, very limited even at 2800 °C. The appearance of the g-value anisotropy was correlated with striking changes in the magnetoresistive effect and the thermoelectric power. Both of these properties exhibited a reversal of sign which may be linked with the significant changes of structure and electronic nature which occurs in the fibres within this region of the heat treatment.

The anisotropy of the g-value was subsequently [71] found to be a good measure of the degree of graphitisation of the fibres. The magnitude of this crystallite anisotropy was consistent with the limited degree of crystallite development characteristic of the non-graphitising nature of these materials. Further studies [72] of the resistivity, magnetoresistance, thermoelectric power and ESR data verified the significance of the processing temperature of 1750 °C reported earlier. These results again indicated a major change in the electronic structure of the fibres in this temperature regime, which appears to correlate with a sudden increase in the degree of graphitisation.

In the fifth paper of the sequence, Robson et al. [73] reported that a linear relationship had been found between the ESR g-value anisotropy, the electrical resistivity (ρ_{300K}), the resistivity ratio (ρ_{77K}/ρ_{300K}), the thermoelectric power and the magnetoresistance of a large number of PAN-based carbon fibre precursors. The correlations also appeared to extend to cellulose-based fibres. A 'graphitic order parameter', defined by the resistivity ratio, correlated well with the crystallite size determined by x-ray crystallography. The ESR g-value anisotropy, however, correlated better with the interlayer spacing, in accordance with theoretical considerations.

Pschenichkin et al. [74] used ESR to identify differences in the structural transformation features of PAN-precursor carbon fibres manufactured by different production technologies and processed at temperatures up to 300 °C. The concentration of paramagnetic centres (PC) for type II fibres increases monotonically as processing temperatures rise to 300 °C, probably due to enhancement of the level of structural homogeneity of the precursor fibre. For type I fibres a maximum PC concentration occurs at around 235 °C in air. The influence of isothermal heating in PAN-fibre carbonisation, on the ESR parameters, was established to lead to acceleration of the process of formation of the polyconjugated system within the fibre.

The Pschenichkin study was continued [75] by investigation of the heat treatment of viscose rayon and PAN fibres in the temperature range 1500–3200 °C, in order to determine the nature of the paramagnetic centres affecting the ESR response. No ESR signal was detected in the temperature range 2400–2600 °C for rayon fibres or 2300–2500 °C for

PAN fibres. Both fibres were known to be 'homogeneously non-graphitising' materials. A similar gap in the ESR signal had previously been reported for non-fibre homogeneously graphitising materials. The absence of signals in these fibres may be connected with the completion of the formation of aromatic structure fragments prior to the onset of heterogeneous graphitisation. Two types of paramagnetic centre were shown to be responsible for the ESR absorption: localised and mobile electrons. The two types occurred in a ratio which varied with the heat-treatment temperature.

Kotosonov et al. [76] used the ESR signal intensity to study the effect of heat-treatment temperature (HTT) for PAN-based carbon fibres. The curve of signal intensity against HTT was compared with analogous data for petroleum coke and phenol–formaldehyde resin, as representatives of graphitisable and non-graphitisable carbons respectively. All three curves were of a similar form, but the fibre line had far higher intensity values in the range 800–1500 °C. This, and the values intermediate between those for coke and resin above 1500 °C, were ascribed to the partial graphitisation of the fibres.

Bright and Singer [77, 78] used ESR to study carbon fibres and demonstrated that preferred orientation, degree of graphitic character, nature of the unpaired spins, impurities and defects can all be interpreted in the ESR spectra. The g-anisotropy was plotted against HTT in the range 1600–3000 °C, but was so large that the spread in g-values broadened the signal orientation. In PAN fibres, which were non-graphitisable, the g-anisotropy saturated at a value equivalent to pitch-based fibres at 2000–2300 °C. The anisotropy of g-values for pitch fibres was much larger than that reported for PAN fibres. Purification of the fibres with chlorine could narrow the linewidth by as much as a factor of five. McClure et al. [79] recently developed a theory for the g-anisotropy in carbon fibres, but a problem appears to exist with the algebraic sign of the anisotropy.

Jones and Singer [80] continued the ESR studies of PAN- and meso-

TABLE 2

Degree of Anisotropy for Computer Simulated ESR Spectra

Parameter	PAN	Pitch	Graphite
$g_1 = g_2$	2.0027	2.0026	2.0026
g_3	2.028	2.035	2.050
Δg	0.0253	0.0324	0.0474

phase pitch-based carbon fibres after heat treatment at 3000 °C. A novel computer program was used to simulate the fibre ESR spectra, based on details such as different degrees of transverse motional averaging, various 'single crystallite' lineshapes and the degree of preferred orientation of crystallites with respect to the fibre axis. Single crystallite ESR parameters were deduced from comparisons with experiments. The degree of anisotropy for the computer-simulated ESR spectra increased in the sequence: PAN-based fibre < pitch-based fibre < single-crystal graphite, as summarised in Table 2.

7.3.3. ESR of Kevlar fibres

Brown and Hodgeman [81] used ESR spectroscopy to study the thermal degradation of Kevlar 49 aramid fibre in the temperature range 350–550 °C. Accumulation of paramagnetic centres occurred above 370 °C in air and above 470 °C in vacuum. Above 520 °C the rate of formation of these radical centres was approximately the same in both environments. Measurement of the g-factors showed that the formation of radical centres in air was associated with a greater amount of chemically bound oxygen atoms than was the case in vacuum. The initial untreated roving showed a weak, slightly asymmetric absorption at $g = 2.0048$ and $\Delta H_{max} = 1.4$ mT. The thermally formed paramagnetic centres showed a strong symmetrical ESR absorption of nearly Gaussian lineshape at $\Delta H_{max} = 0.6$ mT. The signal was stable at room temperature over extended periods and was unaffected by oxygen. The temperature dependence of the ESR signal intensity obeyed the Curie law, indicating that the paramagnetic centres were free radical in nature. The sample heated in air showed an initial decrease in g-factor, levelling off at 2.0030.

Brown and Sandreczki [82] used an X-band ESR spectrometer with homodyne detection and 100 kHz modulation to investigate the free radicals created in Kevlar 49 fibres as a result of stress- and photo-induced macromolecular chain scissions. The signals were recorded on magnetic tape for subsequent field shift, scaling, subtraction, addition and integration. Signal accumulation times of up to 6 h were used (600 scans at 35 s/scan). The fibres were commercial poly(para-phenylene terephthalamide) (PPTA) or aramid polymer made up as 380 denier yarn consisting of ~ 267 filaments with a diameter of 10 μm and 3 turns/metre.

Background spectra in the untreated PPTA fibres indicated the presence of several transition metal ions and of three or more different types of free radical. Mn^{2+} ions were positively identified from the spectra of frozen solutions of the fibre in concentrated sulphuric acid. The other paramagnetic transition metal ions present were Cu^{2+} and possibly Fe^{3+}, Cr^{3+} and Ti^{3+}. The observed anisotropy in the lineshapes indicated that some of the metal ions, together with their first coordinate spheres, were oriented in the crystalline regions of the fibres. Solutions of $2\frac{1}{2}$ w/o and 10 w/o of fibre in sulphuric acid had spectra which consisted of a single intense line at $g = 1.985$, and approximately 18 less-intense lines with a magnetic field spread of 20 mT. The onset of liquid crystalline domain ordering is supposed to occur at 9 w/o Kevlar, but the two spectra did not differ. The anisotropy of lines assigned to the metal ions (due to fine structure, hyperfine structure or g-anisotropy) is 90 mT, whereas the g-anisotropy in the free radical line has field shifts of only 0.3 mT. Stress-induced free radicals were observed in annealed samples of Kevlar 49 fibres that had been fractured in vacuum. The concentration of stress-induced radicals (2×10^{10} per filament) suggested that scission occurred in more weak planes than had been estimated to exist in the fracture surface of the fibre core. These radicals were unstable in air and had some aromatic character associated with their structure. No stress-induced radicals were observed before fracture.

Several different types of free radicals were obtained following UV irradiation of the Kevlar fibres in air (photooxidative radicals) and in vacuum (photodegradative radicals). The photodegradative radicals are identified with:

(a) primary radicals involved with the photo-Fries rearrangement reaction;
(b) secondary radicals formed by hydrogen atom abstraction by the primary radical; and/or
(c) ketyl radicals produced by UV irradiation of the photo-Fries rearrangement product.

The photooxidative radicals are identified with the UV-irradiation products of a peroxide intermediate. The observed anisotropy in the lineshape indicates that both types of radicals are oriented in the crystalline regions of the Kevlar 49 fibres (Fig. 3). The best resolution of the hyperfine structure is observed at 0° with the fibre aligned with the applied Zeeman magnetic field.

Brown et al. [83] used ESR to examine the macromolecular chain scission in the stress rupture of Kevlar 49 fibres. The fibre consists of a 0.1 to 1.0 μm thick skin surrounding a core region which contains bundles of 60 nm diameter cylindrical crystallites oriented in the fibre direction. In the skin region the chain ends are randomly distributed, whereas the inner

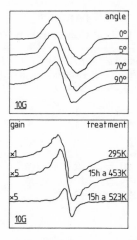

FIG. 3. Top: background spectra of Kevlar 49 fibres with changing orientation. Bottom: ESR spectra of Kevlar 49 fibres after heat treatment (reproduced from Brown, Sandreczki and Morgan [83] by permission of the publishers, Butterworth & Co. (Publishers) Ltd ©, and the University of California Lawrence Livermore National Laboratory, from work performed under the auspices of the US Department of Energy).

core contains clusters of chain-ends leading to periodic transverse weak planes separated by a distance of 220 nm. Some chains traverse the weak planes and maintain the structural integrity. Fracture studies highlighted the weakness of both the transverse planes and the 60 nm diameter intergranular region. Stress radicals were not detected by ESR at stress levels below fracture. After fracture, the number of radicals detected was 50 times that expected from a clean fracture across one transverse fracture plane, equivalent to all the chains in the weak planes for a longitudinal distance of 5 μm either side of the central plane undergoing scission. In practice, both Kevlar filaments and epoxy composites failed in tension by axial splitting over 200–500 μm along the axis. The implication was that only 2–5% of the chains crossing the weak plane were broken. The radicals created were stable in vacuum but not, however, in air, and had some aromatic character in their structure.

Ogo et al. [84] used ESR spectra to examine the concentration of free radicals in fibre-reinforced plastics, generated as a result of the application of shearing deformation under high pressure (mechanochemical reaction). Fremy's salt [$(KSO_3)_2NO$] was used as a reference substance, with ESR g-

TABLE 3
Relative Free Radical Concentrations (from Ref. [84])

Static pressure (kbar)	Shearing angle (degrees)	FRP 1 (woven glass)	FRP2 (chop-strand glass)	FRP3 (woven Kevlar)
20	45	1.00	2.87	19.3
	90	1.51	5.41	30.5
	180	2.36	3.65	13.5 (sic)
30	45	1.53	11.7	19.6
	90	2.63	14.3	39.8
	180	5.47	15.4	47.0
40	180	3.75	—	—
τ_{max}(kg mm^{-2}) at 20 kbar		194	174	28

values between 2.005 and 2.006. The mechanical loading [85] was achieved by compressing two specimens between the two platens of a hydraulic press with an anvil interspersed between them which could be rotated to produce shear stresses. Three groups of composites were tested:

(a) plain weave fabric E-glass fibres in unsaturated polyester resin, FRP1;
(b) chopped strand mat E-glass fibres in unsaturated polyester resin, FRP2;
(c) satin weave cloth Kevlar 49 fibres in epoxy resin matrix: low fibre volume fraction, FRP3.

Table 3 summarises the relative values of free radical concentrations for the various tests, normalised to FRP1 and 20 kbar static pressure sheared to 45°. The free radical concentrations increased with the magnitude of the static pressure and of the shearing angle. The concentration of radicals produced was found to follow the decreasing sequence FRP3 ≫ FRP2 > FRP1. The attenuation of the relative radical concentration descended with time under static loading conditions (30 kbar, 45° shear). The radicals of the Kevlar composite were considerably more stable, but it is unclear whether the change of matrix or of reinforcement is responsible. The Japanese authors attributed the detected free radicals to the rupture of molecular chains within the matrix, which agreed with their conclusions arrived at on the basis of the comparison of τ–θ curves.

7.3.4. ESR of glass fibres

Barbashov *et al.* [86] used both ESR and infrared spectroscopy to investigate structural changes in two materials after fast-electron bombardment (2 MeV), and their effect on the mechanical properties of the laminates. The materials were:

(a) 60 v/o glass-reinforced textolites EDT-10;
(b) EDT-bonded textolite with a lavsan (PET) cloth filler; and
(c) SGM with TS-8/3-250 glass cloth.

Electron spin resonance was used to detect free radicals due to bond scission by the ionising radiation. The form of the spectra did not change with increased radiation dose. A weight-percentage weighted summation of the spectra of the fibre and the matrix resulted in a calculated spectrum which differed from that obtained experimentally. The difference was assumed to be due to chemical interaction between the free radicals and the fibre or matrix during the formation of the interface. A correlation was found between a large change in tensile strength and a greater change in the accumulation of free radicals in SGM relative to EDT-10 after electron bombardment.

Klimanov and Rubakova [87] used ESR and infrared spectroscopies to examine the anisotropy of glass fibres of various compositions, including E-glass. The ESR spectra were recorded after irradiation with gamma-rays (36 Mrad, ^{60}Co, room temperature) on a Varian E3 spectrometer, using a quartz bulb of 4 mm diameter. The filaments were placed in the bulb parallel to each other, and the bulb was then rotated within the apparatus. No anisotropy of the paramagnetic signals was established from iron ions present as admixtures in the E-glass. Anisotropy was observed in some quartz filaments.

TABLE 4

Percentage of Free Electrons in Optic Glass Fibre at Monthly Intervals After X-ray Irradiation (Value Immediately After Irradiation Defined as 100% in Each Case)

Radiation dose (roentgen)	Delay (months)				
	0	2	3	4	5
300	100	43	14	8	4
1000	100	68	26	5	3
2000	100	66	34	7	4

Kato et al. [88] used an X-band (9.64 GHz) ESR spectrometer with a continuously sweeping magnetic field of 3500 to 3275 gauss to examine changes in a 14 µm diameter optical glass fibre arranged in bundles of 12 000 fibres for use as a medical fibrescope. The samples were irradiated with between 5 and 2000 roentgen of X-rays and examined immediately after irradiation and then at monthly intervals. Below 300 R the signals in the ESR absorption curves were barely separable from the characteristic machine noise. The response above 300 R is summarised in Table 4.

Kato et al. concluded that:

(a) after 100 R X-ray exposure changes were detectable in the ESR absorption curves;
(b) distinct ESR absorption curves are produced above 300 R of X-rays;
(c) the shape of the curve above 300 R is proportional to the X-ray dose;
(d) thermal annealing occurs with time, and is more rapid in specimens exposed to low irradiation;
(e) the existence of effects at less than 75 R of X-rays is strongly inferred from data at high X-ray doses.

Cociu et al. [89] used ESR to study the ageing of laminated electrical insulators consisting of mica paper on a delicate glass tissue support impregnated with epoxy resin. The ageing was achieved by cycles of 5 to 10 h at 225 °C followed by cooling to room temperature. Before ageing, the ESR spectrum consisted of a single line at $g = 4.2$, attributed to iron ions. After ageing, an additional line appeared at $g = 2.001 \pm 0.001$, typical of free radicals. The intensity of this new narrow resonance signal increased with the duration of the ageing. The ESR spectra of the mica paper and the glass support were unchanged by the heating, hence the free radicals appeared to be due to degradation of the epoxide. Because the spectra of the composite were recorded at different gains, the intensification of the free radical signal was expressed as the ratio of $I(g = 2.0)$ and $I'(g = 4.2)$, the signal intensities. The increase of free radical concentration was found to rise exponentially towards saturation, with three distinct slopes being distinguished. The changes in slope were claimed to be due to phase transitions and the creation of additional free volume at higher temperature. Eventually an equilibrium was achieved at the high temperature. The dielectric properties of the material deteriorated with the increase in free radical concentration. It was suggested that the lifetimes of electroinsulating transformer laminates should be determined by the saturation level for the free radical concentration.

7.3.5. ESR of epoxy resin

Schaffer [90] used ESR to evaluate epoxy matrix materials intended for use in CFRP structures in space. The spectrometer had an operating frequency of 9.35 GHz, a magnetic field of 3720 ± 120 gauss modulated at 0.63–1.60 gauss and a scan time of 5–10 min. Free radical concentrations were estimated by reference to a free radical standard, DPPH (2,2-diphenyl-1-picryl hydrzyl) in 3M epoxy–amide resin. The epoxy resin under test was a commercial TGDDM/DDS system, which was exposed to irradiation by:

(a) 10–150 Mrad, 0.5 MeV electron radiation at room temperature; or
(b) 5–10 Mrad, 1.33 and 1.77 MeV ^{60}Co gamma irradiation at liquid nitrogen temperature.

From the radical concentration versus dose curve for gamma-irradiated samples a slope of 0.59 radicals/100 eV was obtained. From the radical build-up curve for electron irradiated samples an estimated spur (spur = group of radicals) diameter of 45.4 Å was determined. The decay of radical concentrations was observed at room temperature in both cases. The decay data was found to fit a model which assumed two simultaneous second-order reactions occurring in different zones. The decay constants ranged from 14.2×10^{-21} g/min spin to 1.07×10^{-21} g/min spin for the fast decay species and 0.009×10^{-21} to 0.172×10^{-21} g/min for the slowly decaying species. The results were consistent with various reports of electron microscope, NMR and ESR spin probe evidence of the inhomogenous distribution of regions of high and low crosslink density in cured epoxy resins.

7.4. RAMAN SPECTROSCOPY

7.4.1. The technique [91]

The Raman effect is a phenomenon observed in the scattering of light as it passes through a material, such that the light undergoes a change in frequency and a random alteration in phase. Raman scattering differs in both these respects from Rayleigh and Tyndall scattering in which the scattered light retains the frequency of the incident light and bears a definite phase relationship to it. The intensity of Raman scattering is roughly one-thousandth of the Rayleigh scattering in liquids, and even less in gases. The Raman effect is usually referred to as 'combination scattering' in the Soviet literature.

The development of the laser revived interest in the Raman effect. Laser radiation is intense, polarised and coherent, and highly-collimated small-diameter beams of monochromatic radiation can be produced. Sources throughout the visible spectrum and adjacent regions are available. Continuous wave argon-ion or krypton-ion 1–10 W sources are most often employed, although tunable dye lasers are commonly used in the excitation of resonance Raman scattering (RRS). Raman scattering is analysed by spectroscopic means. The scattering is approximately uniform in all directions and is usually studied at right angles, so that the exciting beam does not interact with the detector.

When the incident beam has a frequency range within the absorption band of the molecule, the radiation may be scattered by two processes: resonance fluorescence or resonance Raman effect. Resonance fluorescence differs from resonance Raman in that the absolute frequencies of the fluorescent spectrum do not shift when the exciting beam frequency is changed, provided that the latter remains within the absorption band. However, the absolute frequencies of the resonance Raman effect shift by exactly the amount of any shift in the exciting frequency, as in the normal Raman effect. The main characteristic of the resonant form as compared with the normal Raman effect is the intensity, which may be two or three orders of magnitude greater. Excellent review articles on resonance Raman spectroscopy (RRS) have been prepared by Behringer [92, 93].

Other special forms of the Raman effect include Hyper-Raman effect and stimulated Raman effect (SRE). A special development of SRE is Coherent Anti-Stokes Raman Spectroscopy (CARS) which uses one fixed frequency and one tunable laser. No reference to the use of these special forms for fibre-reinforced plastics is known at this time.

7.4.2. Applications of the Raman effect in composites

Tuinistra and Koenig [94] used Raman spectroscopy to characterise the surfaces of carbon and graphite fibres. A correlation existed between the Raman spectrum of the fibre surface, reflecting the surface treatment, and the shear strength of the composite. Many graphite fibres have a weak surface layer where the graphite planes are oriented parallel to the surface.

Koenig and Shih [95] used a laser Raman study to confirm that the reversible nature of silanol hydrolysis at the coupling agent to the glass-fibre interface was a dynamic equilibrium with penetrant water molecules competing with the VTES (vinyl triethoxysilane) coupling agent. The composite was E-glass fibre-reinforced polymethylmethacrylate. The intensity of the 788 cm^{-1} SiOSi symmetric stretching line was reversible. After two hours in boiling water, or four months at 100% RH and 38 °C, the line shifted to 783 cm^{-1}, a vinyl silane homopolymer position. Subsequent heat treatment at 110 °C for three days restored the line to 788 cm^{-1}. They concluded that:

(a) an aqueous solution of VTES conventionally applied to the fibres could be shown by laser Raman spectroscopy to be chemically bonded to the glass surface;
(b) Chemical bonding was confirmed by cleaning the fibre with toluene and boiling water; vinyl polysiloxane was still detected on the fibre surface, and enhanced Raman scattering was observed, reflecting the SiO silane bonding;
(c) copolymerisation of the methylmethacrylate and vinylpolysiloxane occurred.

Shih [96] used laser Raman spectroscopy to examine the chemical reaction between a methacryl silane coupling agent on glass fibre and the styrene monomer in unsaturated polyester resin. The styrene spectrum of this sample was identical to that of the homopolymer, but the carbonyl stretching frequency at 1718 cm^{-1} in the unreacted silane was shifted to 1702 cm^{-1} after polymerisation. These results indicated that the styrene was homopolymerised, but with the silane grafted to the chain end. Further, neither methacryl silane nor vinylsilane reacted with the polyester resin in the absence of the styrene monomer.

Koenig et al. [97] used laser Raman spectroscopy to demonstrate the various structures which result from the molecular adsorption of silanol onto a glass surface. There were several lines around 1050–950 cm^{-1} due to the SiO stretching mode of the silanol. Glass microspheres displayed two lines at 980 cm^{-1} for wet samples and at 1005 cm^{-1} for a dry sample. These lines are related to the hydrogen bonding between the silanol and the adsorbed water molecules. An additional line at 992 cm^{-1} was due to hydrogen bonding between adjacent silanols.

Ishida et al. [98] adopted ultraviolet resonance Raman spectroscopy (UV-RRS) for the detection of molecular-thickness layers of a new potential coupling agent for fibreglass composites, as Raman and Fourier-transform infrared spectroscopies had proved insufficiently sensitive. The UV-RRS technique was shown to provide a wider selection of excitation lines than the restricted selection in the visible spectrum. An excitation line at 363.8 nm was found to be a major resonance Raman enhancement for the new coupling agent. Enhancement was 50-fold and 200-fold according to the dissolution medium: KBr and 10^{-4} M benzene respectively. As a surface species the apparent enhancement was much greater. The technique is thus well suited to the study of surfaces, and especially layers of monomolecular thickness or less. The 1545 cm^{-1} line in the silane spectrum, which is sensitive to intermolecular interactions, is also subject to major enhancement.

Galiotis et al. [99] used resonance Raman spectroscopy (RRS) to determine the frequencies of the vibrational modes of the backbone chain in polydiacetylene fibres (PDA), and the dependence of those frequencies on the tensile strain. With a 'reasonably transparent' matrix the resonant enhancement could be obtained from fibres covered by several millimetres of resin. Lap-joints between two PDA fibres cemented with epoxy resin were subjected to a tensile strain. The RRS was undertaken using the 676 nm line of a krypton laser at $\sim$ 5 mW power with a focal spot of 100 μm diameter at the fibre. Data were collected at ambient temperature in the 180° backscatter mode for the vibration at 2085 cm^{-1}, which is associated with the large amplitude of vibration of the triple bond. The Raman frequencies were constant, to within the experimental accuracy, for different positions in a single fibre region. The relationship between frequency and displacement was approximately linear and showed no hysteresis. Similarly, the RRS technique was used to monitor the fibre strains within the lap-joint, where a linear frequency–displacement relationship was again found.

Galiotis et al. [100] have suggested that single-crystal PDA fibres offer considerable potential for use as the reinforcement in strong, stiff, all-polymer fibre-reinforced composites. They have shown that the frequency associated with stretching of the triple bond, as measured by RRS, changes by -20 cm^{-1}/% when the fibres are subjected to tensile deformation. The fibre can thus be used as an internal molecular strain gauge, monitored optically using the laser Raman technique. The strain can be measured to an accuracy of ± 0.05% with a point-to-point resolution of about 100 μm. A model composite system [101] was made consisting of a single single-crystal polydiacetylene fibre in a transparent epoxy resin matrix, and was subjected to longitudinal tensile strain. Resonance Raman spectroscopy was used to determine the strain at all points along the fibre, whilst matrix strain was monitored by conventional techniques. Below 0.5% matrix strain the composite demonstrated Reuss-type behaviour: equal stress in both the fibre and the matrix. At higher strains, the composite exhibited Voigt-type behaviour: increase of matrix strain matched by an equal increase in fibre strain. In this high strain region the simple Cox shear-lag model provides a qualitative description which explains most of the observed results.

Young et al. [102] have extended the use of the RRS technique to the measurement of the distribution of fibre strains in composites containing bundles of PDA fibres in both an epoxy matrix and a fibreglass/epoxy composite. The deformation of epoxy composites with a high volume fraction of single-crystal PDA fibres was investigated in both tension and compression.

Young et al. [102a] have recently extended their work to include the Raman frequency shift with strain in Kevlar fibres and report a value of 5 cm^{-1}/%. Using laser beams that can be focused down to less than 2 μm has allowed point-to-point variation of strain to be determined along individual fibres in polymer matrix composites to a high degree of precision. Measurements have been made of the dependence of transfer length upon fibre diameter. Preliminary studies have been made of the effect of surface pre-treatment upon stress transfer and of the interaction between cracks and reinforcing fibres.

7.5. FOURIER-TRANSFORM INFRARED SPECTROSCOPY (FTIR)

7.5.1. The technique [103]

Infrared spectroscopy is widely used as a technique to identify organic substances, including plastics. All chemical compounds have characteristic intramolecular vibratory motions which can absorb incident radiant energy if that energy is sufficient to increase the vibrational motions of the atoms. In organic molecules the vibrational motions of the substituent groups within the molecules coincide with the infrared region of the electromagnetic spectrum. Traditionally, infrared spectroscopy was achieved by the use of gratings or prisms to scan through the frequency range transmitted by the sample and thence plot a graph of energy incident on the detector against frequency or wave number (in cm^{-1}). Because the bulk of the available energy does not fall on the open slits, the sensitivity of this method is very limited. This energy limitation is particularly severe for polymer analysis, where the absorption bands are generally broad and weak.

Fourier-transform infrared spectroscopy (FTIR) was developed to improve the sensitivity relative to that of the traditional method, by allowing the examination of all the transmitted energy all of the time. The technique uses the Michelson interferometer in which the equally split infrared beam is reflected from a stationary plane mirror or a movable mirror. After reflection the two perpendicular beams are reunited at the beam splitter and travel on to the detector. The reflected beams combine either constructively or destructively, depending on the relationship between their path difference and the wavelength of the radiation. When the path distance from the beam splitter to each of the two mirrors is identical, all wavelengths of the radiation add coherently to produce a maximum flux at the detector, known as the 'centre burst'. Displacement of the moveable mirror changes the length of that arm of the interferometer, and the resulting difference in path length causes each wavelength of source radiation to destructively interfere with itself at the beam splitter. The resulting flux at the detector, being the sum of the individual wavelengths, rapidly decreases with mirror displacement. An interferogram is obtained at the detector by sampling the flux while moving the mirror. For greater accuracy the signals from a number of scans can be averaged to reduce signal–noise errors. The amplitude of the light incident upon the detector at any specific frequency can be derived from the interferogram by the use of Fourier transform theory. The use of modern computers and the fast Fourier transform (FFT) algorithm [104, 105] allows the infrared spectrum to be calculated in less than one second, while the moving mirror is returning to the centre burst position.

FTIR has a number of advantages over traditional infrared spectroscopy. Theoretically, an FTIR spectrometer can acquire a 0–4000 cm spectrum with 1 cm resolution, with the same signal-to-noise ratio, 4000 times faster than a dispersive instrument. Alternatively, if the same time is taken to acquire both spectra there will be a 63-fold increase in the signal-to-noise ratio on the FTIR instrument. The short time required (1.5 s) to obtain a complete FTIR spectrum allows reactions such as the curing of epoxy resins to be followed closely. FTIR is internally calibrated against the laser, whereas the traditional instrument is subject to drift. Three different traditional instruments are required to study near-, mid- and far-infrared spectra, but a single FTIR instrument can be easily converted to study the full frequency range.

FTIR may be operated in a number of modes: transmission, diffuse reflectance (DR), internal reflection spectroscopy (IRS), multiple internal reflection (MIR) and attenuated total reflection (ATR).

7.5.2. FTIR of epoxy resin composites during cure

Liao and Koenig have recently reviewed the application of FTIR spectroscopy to the study of fibre resin composites [103], and the reader is referred to that work for a more comprehensive treatment than follows here. The assignment of characteristic infrared absorptions for common epoxy resins can be found in Lee and Neville [106], while Antoon [107] has made a comprehensive assignment of infrared and Raman bands in the EPON 828/nadic methyl anhydride system. Pearce et al. [108] has studied the spectra of three epoxy resins (DGEBA, DGEPP and DGEPF) in the original state, and subject to thermal degradation, thermooxidative degradation and photooxidative degradation. Summary tables of the band assignments made by Antoon and by Pearce can be found in Ref. 103.

Studies of the curing of epoxy resins by infrared spectroscopy are usually based on the absorption intensity of the epoxy, anhydride and hydroxyl functional groups which appear at 913, 1858 and 345 cm^{-1} respectively. The near-infrared spectrum of the epoxide functional group may also be used. It is very difficult to measure the infrared spectrum of epoxy matrix fibreglass composites with dispersive spectrometers because the glass fibres have a very strong absorption in the mid-infrared region.

Buckley and Roylance [109] used fast-scanning (1 s) FTIR to study the kinetics of a sterically hindered amine-cured epoxy resin system. Variation in the epoxide absorbance due to difference in the specimen thickness were eliminated by the use of an internal reference peak at 1510 cm^{-1} due to the phenyl groups. If the absorbance is represented by $A_{x,y}$ where x is the specimen epoxide or reference absorbance and y is the time since commencement of the reaction, then the fraction of unreacted epoxide at time t will be given by:

$$f_{915} = \frac{A_{915,t}}{A_{1510,t}} \times \frac{A_{1510,0}}{A_{915,0}} \qquad (11)$$

Sprouse et al. [110] used FTIR to monitor the extent of cure in fibre-reinforced epoxy composites by internal reflectance spectroscopy. Infrared spectra were recorded at short time intervals throughout the cure cycle and compared with the spectra of thin films of neat resin, from the batch used in composite fabrication, sandwiched between salt plates.

Antoon et al. [111] demonstrated the usefulness of FTIR difference spectra, in combination with a least-square curve-fitting program for improved precision, for the investigation of the composition of neat epoxy resin, hardner, catalyst and cured resin and the degree of cross-linking in the latter.

Pater and Scola [112] used FTIR to study the contamination of composite surfaces during fabrication as a result of the transfer of silicone release agent. Four standard calibration points were established on the absorbance curves (1273, 1070, 1025 and 775 cm^{-1}) for the quantitative analysis of the silicone contaminants. The method required only small amounts of material and could provide both qualitative and quantitative information in a single analysis. Several factors were found to influence resin transfer, including type of mould, mould surface treatment, type of release agent and the reinforcement fibres. Ease of transfer of silicone to the composite decreased in the order graphite > Kevlar > glass. Silicone penetrated into the bulk of the composite, probably by solution in the uncured resin and diffusion during the cure cycle. The shear strength of a fibreglass composite with deliberately included release agent (0–20 mg/cm^2) declined with increasing silicone content.

Barbashov et al. [86] used conventional infrared spectroscopy to study the changes in the chemical structure of epoxy-bonded reinforced plastics under the influence of electron bombardment. Three significant absorptions were considered: the 852 cm^{-1} band (oscillations of the epoxy ring), the 915 cm^{-1} band (aromatic ring additions or unsaturated carbon–carbon bonds) and the 708 cm^{-1} band (strain oscillations of the aromatic ring). The logarithm of incident radiation over transmitted radiation (the optical density) was plotted against the radiation dose. The initial optical density represented the maximum level and a characteristic absorption minimum occurred at 150 Mrad, followed by a monotonic decrease beginning around 350 Mrad. This characteristic is assumed to be due to two opposing effects: crosslinking and radiation-induced degradation.

Walker et al. [113] used traditional transmission infrared measurements to study the void content of fibreglass–epoxy NOL-rings. A transmission band at around 1.78 μm was found to be convenient for all samples. The transmission of the glass fibres ceased at 2.5 μm. A 15% transmission of the infrared radiation was achieved in the composites with the lowest void content, although the extent of this transmission was dependent on the fibre volume fraction. Scattering by the voids appeared to be the dominant mechanism for the dissipation of the infrared in samples with high void content. The weight percentage of glass fibres could be estimated from the difference in absorbance (ΔA) between the two transmission maxima (2.15 μm and 1.96 μm) such that the volume fraction of the glass fibres, V_f, was given by:

$$AV_f = 0.385 \quad (12)$$

7.5.3. Epoxy composite degradation monitored by FTIR

George et al. [114] used FTIR to investigate the photoprotection of the surface resin in a glassfibre-reinforced composite. A single-pass internal reflectance attachment was used. A strong ester carbonyl group at 1735 cm^{-1} was used as an index of the photooxidation due to exposure to the sunlamp. This band can be used to measure photooxidation rates in a number of different epoxy resins. The observed oxidation rates for epoxy–novolac were eight times those for bisphenol-A epoxy. The high photooxidation rate of the cured epoxy–novolac is related to the cure process, in which a chromophore is formed unless the cure is carried out under vacuum. Similarly, a strong carbonyl group absorption due to oxidation appeared in the IR spectrum during cure, unless carried out *in vacuo*. The weathering stability of an epoxy resin would thus appear to depend on the conditions during cure.

Pearce *et al.* [108] used the subtraction of absorbance spectra to compare the stability of functional groups within the DGEBA epoxy resin before and after thermal degradation. Some of the absorption bands in the cured resin decreased in intensity after heating, while new bands also appeared. From a sequence of difference spectra the order of functional group stability was established to be: total methyl group ~ total benzene group > methylene > *p*-phenylene > ether > isopropylidene.

Young *et al.* [115] used diffuse reflectance (DR) FTIR spectroscopy to study cured graphite fibre composites. In each case significant changes in molecular structure of the resin were observed, which were correlated with previously observed changes in material properties. In the graphite/epoxy composites the phenyl and sulphone bands were not significantly changed by ageing at 121 °C. Oxidation of the methylene groups to carbonyl groups was evident as new bands between 1700 and 1650 cm^{-1}. The graphite/polysulphone was exposed to 10 Grad of electron bombardment, which resulted in a decrease in intensity or disappearance of the sulphone group bands (1409, 1294 and 1150 cm^{-1}). In the linear polyimide composite no shifts in frequency or changes in band intensity were observed after 25 000 hours at 232 °C. After the same time at 288 °C there was a marked decrease in intensity, suggesting resin loss from the oxidation of chain ends. At this point the tensile strength had decreased significantly and the appearance of a very intense band centred on 1105 cm^{-1} was noted. In the addition polyimide composite, degradation proceeded by oxidation of the CH_2 groups of the methylene–dianiline portion of the polymer (new bands at 1667 and 931 cm^{-1} after 15 000 hours at 232 °C in air). The 1599 cm^{-1} phenyl band increased in intensity as the 1511 cm^{-1} phenyl group decreased in intensity, echoing the extension of the conjugation to the adjacent newly formed carbonyl group. The imide vibration bands moved apart, from 1716 to 1724 cm^{-1} and from 1371 to 1361 cm^{-1}. This is believed to be due to constraint of the imide ring vibration by additional curing or crosslinking, such as would occur upon oxidation of the norbornene aliphatic ring followed by maleimide crosslink formation. This DR-FTIR correlates well with the increase in T_g, weight loss and decrease of flexural strength.

Fuller et al. [24, 25] used NMR (q.v.) and infrared techniques to study the absorption of water by TGDDM/DDS epoxy resin. Specimens were soaked in either H_2O or D_2O (deuterated water). The spectrum of samples soaked in H_2O and dried in the IR beam was the same as that of the original dry samples. The spectrum of samples soaked in D_2O showed a shift from 3300 to 2600 cm^{-1}. It was hypothesised that hydrogen exchange was taking place between the protons on the polymer backbone chain and the deuterons from the heavy water. The frequency shift was consistent with expectations as the deuteron has twice the mass of the proton, and the frequency is inversely proportional to the square root of the reduced mass. The frequency of the band (3300 cm^{-1}) is consistent with a hydroxyl stretching frequency. Exchange may of course also occur with unreacted curing agent.

Ishida and Koenig [116] have reviewed the reinforcement mechanism of fibreglass-reinforced plastics under wet conditions, although the majority of that paper is concentrated on effects at the fibre/resin interface (see Section 7.5.5). Antoon [107] studied the hygrothermal stability of glass-fibre-reinforced epoxy composites with respect to the glass reinforcement, the interface and the matrix. FTIR spectroscopy was utilised to clarify both the chemical composition of the matrix and the physical and chemical effects of moisture exposure on the anhydride-cured resin. Digital subtraction of the spectra obtained during copolymerisation indicated a need for several modifications to current thinking about the reaction mechanism. In addition the results showed that the kinetics are diffusion-controlled with no achievement of steady-state concentrations of the active centres. The composition and extent of cure could be rapidly calculated regardless of the filler concentrations. Such calculations were aided by a 'factor analysis' procedure which determined the number of independently absorbing species in the composite system. The composition of any composite could then be determined by fitting the spectra of those species to the composite spectra by a least-squares criterion. Both the reversible and irreversible effects of moisture on the epoxy matrix were investigated. The diffusion of water vapour into the resin perturbed the vibrational modes of both the water and the resin, so as to imply the formation of hydrogen bonds between the water molecules and polar moieties in the resin. The effect was completely reversible. Long-term exposure of the epoxy matrix to liquid water caused the hydrolysis and leaching of unreacted anhydride molecules. Hydrolysis of the ester linkages in unstressed resin was only significant in a highly alkaline medium. However, the application of a high tensile stress to the matrix dramatically accelerated hydrolytic attack in films exposed to alkaline media or neutral water. The mechanochemical degradation was modelled by an exponential dependence of the hydrolysis rate on the applied stress. Incorporation of E-glass or alumina fillers in the matrix caused acceleration of the hydrolytic attack, and this was attributed to the residual stresses in the matrix near the filler surface.

7.5.4. Measurement of fibre stress by infrared spectroscopy

Regel' et al. [117] used infrared spectroscopy to measure the fibre stress in a unidirectional polypropylene fibre-reinforced low-density polyethylene. The magnitude of the stress in the fibre was estimated from the frequency shift of the 1168 cm^{-1} peak, which corresponds to oscillations of the polypropylene molecules in the configuration of an isotactic 12-unit monomer helix. The matrix has no intrinsic absorption within this range. The magnitude of the frequency shift, Δv, was found to be proportional to the mean stress σ in the specimen with a proportionality constant α of 0.1:

$$\Delta v = \alpha \sigma \qquad (13)$$

The results are summarised in Table 5, where all stresses are quoted in kgf/mm^2. The disparity between the stresses predicted by infrared shift and rule of mixtures was assumed to be due to matrix hardening.

TABLE 5

Frequency shift of peak (cm^{-1})	Fibre stress from IR	Fibre stress from rule of mixtures	Composite stress
0.6	6	11	3.2
1.1	11	22	6.4
1.7	17	34	9.9
2.0	20	45	13.0
2.5	25	56	16.3

7.5.5. FTIR of the fibre/matrix interface

Fourier-transform infrared spectroscopy has been used extensively at the Case Western Reserve University for the study of coupling agents on fibre reinforcements, principally glass fibre. The principal coupling agents studied were:

Vinyltriethoxysilane (VTES)	[118, 119, 139]
Vinyltrimethoxysilane (VTMS or VS)	[120, 121, 123, 125]
γ-methacryloxypropyltrimethoxysilane (MPS)	[120, 125, 133–136, 138, 139]
Cyclohexyltrimethoxysilane (CS)	[121, 125, 126]
γ-aminopropyltriethoxysilane (APS)	[122, 124, 127–131, 136–140]
γ-aminopropyltrimethoxysilane	[124]
n-methylaminopropyltrimethoxysilane (MAPS)	[127–129, 132]
n-2-aminoethyl-3-aminopropyltrimethoxysilane (AAPS)	[129]
n-2-aminoethyl-3-aminopropyltriethoxysilane	[137]
Trimethoxysilylpropyldiethylenetriamine	[137]

The majority of the above tests were carried out on finely ground fibres included in potassium bromide (KBr) pellets. In one paper [120] the E-glass fibre was coated with either MPS or VS coupling agents, ground and mixed with a polystyrene solution containing 0.5 w/o cobalt naphthanate cocatalyst and 1 w/o MEKP. The final composition was 50% E-glass, 24.6% polystyrene and 24.6% styrene monomer. The mixture was cast between KBr windows and polymerised in the spectrometer at 50 °C for 1 h then for an additional hour at 80 °C. The spectra were recorded at room temperature. Difference spectra were recorded as well as the conventional spectra; for example, in one case the spectra were recorded and subtracted for E-glass fibre mixed with the styrene–polystyrene before and after polymerisation. VS and MPS were shown to copolymerise with the styrene monomer on the glass surface, demonstrating the covalent bonding at the coupling agent/matrix interface. The silanes existed as multilayers on the fibres. The MPS interphase was completely polymerised with the styrene, but the VS remained largely unreacted. The MPS underwent oxygen-induced homopolymerisation above 110 °C. Untreated E-glass powder was found to inhibit the polymerisation of polyester resin.

Chiang et al. [127, 129, 132] studied APS, MAPS and AAPS on E-glass fibres by FTIR spectroscopy, including KBr pellets of the composite glassfibre-reinforced epoxy/nadic methyl anhydride resin. The structure of the silane-to-resin interface was found to be composed of copolymers of the epoxy resin with the organofunctional groups on the silane. The number of interfacial bonds formed was dependent on the extent of the coupling agent on the fibreglass and on the reaction conditions. The silane was found to induce additional esterification. FTIR suggested that there is a gradient in the degree of cure induced by the coupling agent acting as an initiator. The properties of this resin interphase were different from the properties of the bulk resin, with an increase in the curing density of between 5 and 10% near the fibre relative to that in the bulk resin.

The infrared absorption decreased relative to the baseline, with consumption of epoxy (914 cm^{-1}), anhydride (1780 cm^{-1}) and amine (3290 cm^{-1}) as the composite was cured. Increases in absorption appeared with the formation of ester (1744 cm^{-1}), amide (1640 cm^{-1}) and ester (1183 and 1098 cm^{-1}). The nadic methyl anhydride (NMA) was found to react with both APS and MAPS coupling agents, with the secondary amino-silane (MAPS) having the higher reactivity and actually catalysing the copolymerisation reaction even in the absence of catalyst. No reaction occurred between the epoxy resin and APS below 200 °C, but above 300 °C the reaction and degradation of the resin occurred simultaneously. Copolymerisation between MAPS and epoxy resin occurred at 150 °C.

Kamenskii et al. [141] used conventional infrared transmission and multiple internal reflectance spectroscopy to study the interface between PHA (polyheteroarylene) fibres and an epoxy resin matrix. Two types of

hydrogen bonding were identified within these fibres. An intramolecular hydrogen bond can occur between the carbonyl and amine groups adjacent to one another on the polymer backbone. An intermolecular hydrogen bond can occur between the same groups on adjacent chains. For such aramid fibres with a nominal diameter of 15 μm, indirect data from electron microscopy suggested that the binder could diffuse into the fibre to a depth of a 1–2 μm. The correlation of the infrared spectra with weight percentage of the reinforcement was not an additive sum of the individual components, suggesting that there was a direct physicochemical interaction between the fibre and the matrix. The authors concluded that the resin could penetrate the interfibrillar space and lead to additional reinforcement of the fibre structure, if the correct thermal treatment was followed.

7.6. ULTRAVIOLET SPECTROSCOPY

Ultraviolet spectroscopy can be invaluable for investigating the structures of unsaturated organic molecules [142]. All organic molecules absorb radiation, but only those containing double bonds and unbonded electron pairs absorb in the most accessible region of the ultraviolet spectrum (190–400 nm). The absorption of this UV radiation produces an excited molecule in which one of the electrons of the carbon–carbon double bond or of the carbonyl (C = O) group is promoted to a higher energy level. Ultraviolet data are usually presented in terms of the wavelength of the maximum absorption, λ_{max} (measured in nm or Å), and the intensity, ε (molecular extinction coefficient), at that wavelength. Values of λ_{max} are a measure of the energy required to effect transitions to excited states, and hence indicate the relative stability of the normal and excited states. Radiation at longer wavelengths provides less energy. Compounds containing isolated double bonds absorb in the range 162 to 190 nm, whilst conjugated molecules (those containing alternating single and double bonds) absorb above 210 nm. Extension of the conjugated system intensifies the absorption peak and shifts it further to higher wavelengths, towards the visible spectrum. Typical values of λ_{max} and ε are presented in Table 6.

A Chinese group [143] used UV spectroscopy and silica gel layer analysis (SGLA) to study the reinforcement mechanism at the interface between the glass fibre and the coupling agent. The silane/glass interface was identified by three layers: the physical adsorption layer, the chemical adsorption layer and the chemically bonded layer. The latter two layers made the major contribution to the reinforcing effect as determined by laminal peel and flexural strength tests. The reinforcement effect was more evident when reaction could occur between the resin and the coupling agent. A detained factor, R_f, was introduced which 'can detect if the chemical reaction occurred on the interface or not, this is closely related to the reinforcing effect'. A silica gel was ground in water or 5% water solution of the silane, coated onto a plate and dried. Resin was dropped onto the

TABLE 6
Wavelengths of Maximum Absorption and Molecular Extinction Coefficients in the Ultraviolet Spectroscopy of Simple Organic Molecules [142]

Molecular structure	Name		λ_{max} (nm)	ε
Aliphatic compounds				
$CH_2 = CH\ CH_3$	propene		175	12 500
$CH_2 = CH\ CH_2\ CH_2\ CH = CH_2$	hexa-1,5-diene		185	20 000
$CH_2 = CH\ CH = CH_2$	buta-1,3-diene		217	21 000
$CH_3\ CH = CH\ CHO$	but-2-enal		217	15 650
$CH_2 = CH\ CH = CH\ CH = CH_2$	hexa-1,3,5-triene		257	34 700
Aromatic compounds				
C_6H_6	benzene	(first band)	198	8000
		(second band)	255	230
$C_6H_5NH_2$	aniline	(first band)	230	13 000
		(second band)	280	1430
$C_6H_5CH = CH_2$	styrene	(first band)	244	12 000
		(second band)	282	450

surface and caused to spread by a 1:1 volume mixture of acetone and benzene. The plate was developed in $K_2Cr_2O_7$–H_2SO_4. The value of R_f is derived from:

$$R_f = \frac{\text{Distance moved by the resin spot}}{\text{Distance of front spreading agent edge}}. \tag{14}$$

7.7. LUMINESCENCE

7.7.1. Introduction [144]

The term luminescence was defined by Wideman in 1888 as 'all phenomena of light not solely conditioned by the rise in temperature'. It is now more widely defined as a characteristic non-thermal emission of electromagnetic radiation by a material upon some form of excitation. Luminescence can be rigidly divided into 'fluorescence', having a persistence of less than 10^{-8} s, and 'phosphorescence' with an afterglow of greater than 10^{-8} s. The luminescence process can be divided into three stages:

> absorption of energy;
> excitation; and
> emission of energy.

The emitted energy is, of necessity, of lower magnitude than the absorbed energy. Several types of luminescence have been identified:

bioluminescence	excited by biochemical reactions
cathodoluminescence	cathode rays
chemiluminescence	chemical reactions
electroluminescence	electric fields
mechanoluminescence	mechanical stress
photoluminescence	light
triboluminescence	(synonym for mechanoluminescence)

7.7.2. Chemiluminescence (CL)

George and Pinkerton [145] examined the nonstationary chemiluminescence of organic materials by measuring the weak light emitted when the environment was perturbed by mechanical stress or by UV irradiation or by substitution of an oxidising gas in place of the inert atmosphere. In each case free radicals were created by the disturbance and there was a characteristic rate at which the material attained a steady state in the new environment. The rates of these reactions were very sensitive to the presence of trace impurities, initiators, stabilisers and the microenvironment of the material. A particular advantage of nonstationary CL was the ability to obtain rate parameters at very low extents of oxidation. Assessment was thus rapid with potential for use as a nondestructive monitoring technique. The experiments included oxidation of epoxy resins and thermal/stress CL of nylon fibres, PET fibres and composite materials:

(a) $0°/90°/0°$ three-ply glass fibre in an epoxy–novolac; and
(b) crossplied S2-glass fibres in epoxy resin.

The composites were observed during transverse matrix cracking. When composite (a) was introduced to the CL apparatus there was a high count rate followed by a slow decay over 1 hour. This was found to be due to initiation by the fluorescent room-lights. After low intensity UV irradiation in air, the decay followed second-order kinetics at 112 °C with $\tau = 3.95$ min. The lifetime of these radicals should be a sensitive index of the crosslink density of the composite, through the dependence on the diffusion-controlled termination rate constant. The resin of composite (b) showed nonstationary stress chemiluminescence (SCL) after stressing, with the rate of decay being affected by the extent of cure of the resin and by the presence of moisture. Experiments were performed in single and multiple plies of the composite. No SCL was obtained in axially loaded single-ply samples until failure of the composite, when several bright flashes occurred. In a 0°/90° two-ply composite, strong bursts of CL occurred above a critical strain corresponding to the 'knee' in the stress–strain curve. Removal of the stress was followed by a slow decay of the SCL. Upon stress-cycling of the specimen, successive intensive bursts of light (accompanied by acoustic emission) were followed by a slow decay. This was associated with the formation of transverse-ply cracking. Radical lifetimes of 2.05 min were obtained at 60 °C.

Wolf et al. [146, 147] used chemiluminescence to examine the thermal ageing of polymers (TGDDM/DDS epoxy) and their composites. The chemiluminescence peak intensity which occurred at a temperature change was found to be an inverse function of the ageing time for a given set of environmental parameters. The cause of the CL was assigned to the formation of propenal as a decomposition product. Samples exposed to dry environments exhibited considerably smaller slopes than those exposed to a humid environment. Two mechanisms were suspected of increasing the CL rate:

(a) wet samples aged faster and suffered more bond cleavage; and
(b) oxygen was more soluble in the wet resin.

Although the peak intensity decreased appreciably during the ageing process, the activation energies for the production of CL from each thermal transition were essentially unchanged during ageing, indicating that there was no significant change in the ageing mechanism. The maximum CL peak intensity occurred at the 50°–60 °C transition. Chemiluminescence did not increase much with temperature. After ageing for 200–300 h the chemiluminescence intensity was essentially equal to that of the background. The intercept of the CL/ageing–time graph was found to be independent of ageing conditions, as would be expected for a fresh sample. The slope of the graph was a strong function of both the ageing temperature and the sample moisture content.

7.7.3. Mechanoluminescence (ML)

Rickards et al. [148, 149] monitored the mechanoluminescence (ML) during the failure of glass–tape composites with volume fractions in the range 57–67% and reinforcement angles of 0°, ±10°, ±20°, ±30° and ±45°. During uniaxial tension of a ±45° specimen, mechanoluminescence began at approximately the same time as the stress–strain curve became non-linear, and the photoemission (measured in arbitrary units of light emission: AULE, with no study of spectral distribution) reached 920

AULE. For tension of a uniaxial specimen, the ML began at a slightly higher strain (1.1% against 0.9%) but the stress–strain curve was linear to ~ 2.4% strain, equivalent to damage as detected by ML commencing at ~ 58% of the failure strain.

Upitis et al. [150] used mechanoluminescence to monitor fibreglass tubes under internal pressure and tensile loading. Pressure was used to determine E_{22} and v_{21}, while tension was used for E_{11} and v_{12}. The composites tested were:

(a) Steklonit' VM1 fibreglass in EDT-10 epoxy resin, wet filament-wound into oblique/transverse/longitudinal tubes;
(b) spiral-wound fibreglass in PN-1 polyester resin.
(c) RBMN-10-1680 Steklonit' ropes.

The photoreceiver had an operating range of 300–830 nm with a peak at 400–440 nm. ML intensity was measured by photon counting. Tensile loading of composite tubes from group (a) produced perceptible ML between 70% and 82% of the ultimate tensile stress, with a photoemission count of 100–170 AULE. The ML began in a clearly nonlinear part of the stress–strain curve (72% UTS or 42% ε'). In a similar test of a tube with defects the total AULE count was 300–400. Tensile loading of tubes of type B wound at $\pm 70°$ resulted in mechanoluminescence being first observed at between 80% and 90% of UTS, with a total of 800–900 AULE. The mechanoluminescence was found to increase with the nonuniformity of the material. Under pressure loading tubes (a) exhibited ML at 50–70% of limiting strength, while tubes (b) showed ML at 30–50% of limiting strength. The AULE count for tubes (b) was two or three times greater than for tubes (a). Tubes (b) were also tested in compression, when ML occurred at 85–90% of the compressive strength. The total AULE count was an order less in compression than in tension.

Krauya et al. [151] tested fibreglass–plastics to destruction whilst monitoring mechanoluminescence. Unidirectional dumb-bells and longitudinal–transverse wound tubular specimens were tested in tension, compression and torsion and combinations thereof. The ML intensity was measured by photon counting, with no study of spectral distribution, and with pulse integration every 10 ms. During uniaxial tension the ML began at ~ 59% UTS. Two stages of ML were observed, firstly at 5–20 AULE s^{-1}, then at ~ 1000 AULE s^{-1}. The first low-rate pulses higher than the dark-radiation background of the equipment were formed by microdestructions of the matrix. The later high-rate emission accompanied cracking along the reinforcement fibres and pull-out of long sections of fibre from the matrix. Pullout was accompanied by a 'powerful splash' of photoemission.

Teters et al. [152, 153] studied rectangular flat samples of unidirectional GFRP and OFRP (organic fibre-reinforced plastics) tested in uniaxial tension, and helically-wound $\pm 45°$ tubular samples and crossply tubular GFRP samples tested singly or in combinations of tension, compression and internal pressure. Photoemission was again measured in AULE with no study of spectral distribution. For an epoxy binder, the mechanoluminescence started at 35% UTS and reached 1485 ± 366 AULE. For glass fibres, the ML began at maximum load and reached 236 ± 32 AULE. For axial GFRP, the ML reached $21\,618 \pm 2002$, indicating that rupture of the interface was a dominant cause of ML. Approximately 50% of the AULE were absorbed by a thickness of 337 μm of unidirectional GFRP. The progress of photoemission in this GFRP started with an initial low intensity period (5–20 AULE s^{-1}) until 59% UTS. Powerful flashes of photoemission followed, accompanied by sample damage through longitudinal cracking. These flashes could be observed visually once one was accustomed to the darkness, and were monitored at $\sim$ 1000 AULE per flashup. Below 75% UTS individual light flashes were observed at different sites in the samples. Beyond 75% UTS long lines of light emission, parallel to the fibres, were observed over the entire sample area.

In the unidirectional OFRP, the ML started at 36% UTS with photoemission gradually accumulating. The total power was an order of magnitude less than for GFRP, at 2000–2500 AULE. In the testing of the OFRP with a load history, the photoemission started later and proceeded in a changed manner (Fig. 4). In crossplied tubular samples, the ML began at 12%, 21%, 70% and 96% respectively for tension, internal pressure, shear and compression. In helically wound tubular samples, the ML commenced at 51%, 35% and 55% for tension, internal pressure and compression respectively. The authors suggested that similar mechanisms of failure occurred in tensile and compressive loading, from the similarity of the % UTS at which ML began.

Menzheres et al. [154] investigated the changes in packing density of filled epoxide compositions during hardening by a molecular probe method. Ground glass fibres and Aerosil with a variety of surface treatments were used as fillers either singly or in combination, after their background luminescence had been checked. The resin was ÉD–20 epoxide cured with PEPA (polyethylenepolyamine). Anthracene was introduced as a molecular probe in order to ascertain the average density of the composition during network formation, as it does not undergo chemical

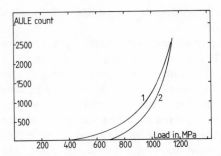

FIG. 4. Relationship between mechanoluminescence and tensile stress in organic fibre-reinforced plastic for virgin material (1) and material with a prior load history (2) (from Teters *et al.* [153], reproduced from *Mechanics of Composite Materials* with the permission of Plenum Publishing Corp.).

changes and the spectrum strictly retains its characteristic electron vibrational structure. Changes in the density of the material during cure and crosslinking result in shifts of the spectra of the probe molecules. The characteristic line in the spectrum of unfilled uncured resin was 1285 cm^{-1}, which was used as a reference line. The line was shifted to 1300–1350 cm^{-1} in the various compositions. The largest spectral shifts of ~ 65 cm^{-1} corresponded to an average increase in the packing density of around 3% relative to the standard reference. For all compositions with glass fibres, the structural changes passed through extremal values with time. The average packing densities for Aerosil/fibreglass mixtures was not predicted by the rule of mixtures.

7.8. FRACTO-EMISSION (FE)

The emission of charged particles, neutral particles and photons has been observed both during and following fracture. Charged particle emissions are primarily electron emission (EE) and positive ion emission (PIE). Excited and ground-state neutral particles are termed neutral emission (NE). Photon emission (phE) is more frequently referred to as mechanoluminescence or triboluminescence. Radio emissions (RE) have also been detected at fracture. These various types of emission share some common features, indicating that they are all produced by common mechanisms. All forms of such emissions accompanying fracture are grouped together under the designation 'fracto-emission', abbreviated to FE.

Fracto-emission is essentially caused by the deposition of a high concentration of energy into a small volume of material during crack propagation. For a short time period (microseconds or less) this can result in:

(a) localised production of heat;
(b) creation of excitations and defects in the material;
(c) production of free radicals, ions and electrons at the crack surface;
(d) emission of particles (electrons, ions and neutral particles) into the gas phase;
(e) separation of charges on crack walls, with intense electric fields resulting in insulators; and
(f) production of acoustic waves (see Chapter 2).

The interaction of excited and reactive species at a surface can readily produce electrons, free ions and photons via Auger deexcitation, stimulated desorption, chemi-emission and chemiluminescence, particularly on a highly reactive surface such as a freshly created crack wall.

Dickinson *et al.* [155–164] have observed that single fibres and pure epoxy resin individually produce electron emission with a simple decay curve and time constants of a few microseconds. The emission curves at fracture of fibre-reinforced epoxy strands made of the same components were found to have emission curves which differed considerably from the pure materials. The time scale for the composites was very long when compared to that of the pure materials. A rapid rise in emissions preceded failure, and the decay exhibited a complex time dependence, lasting whole seconds, rather than microseconds as in pure materials.

Events were detected prior to failure which had decay constants associated with the fracture of the pure samples. The delamination or interfacial failure between the filaments and the resin was believed to be responsible for the major EE component of the slow decay emission. The EE from E-glass and S-glass composites at fracture were relatively smooth curves of counts against time, but the graphite fibre strands in epoxy resin exhibited intense and very erratic EE curves. This erratic behaviour may be due to discharge of the surface by the electrically conducting carbon fibres [155, 157].

The PIE curves for all three composites following rupture were relatively smooth. Comparisons of the EE and PIE revealed that the total emission was almost identical over several samples (Fig. 5). When the emissions from different samples were normalised, it was found that the two curves were indistinguishable within the fluctuations of the observed particle

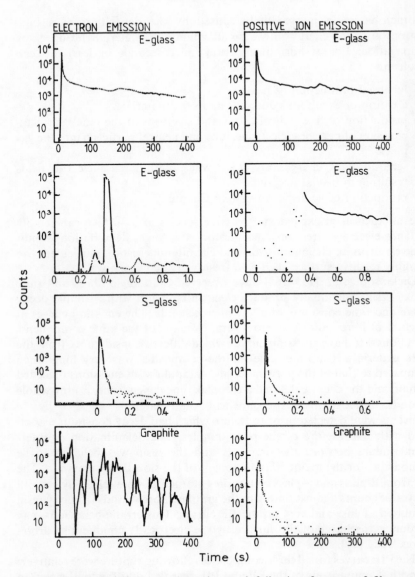

FIG. 5. Fracto-emission accompanying and following fracture of fibre–epoxy strands (from Dickinson, Donaldson and Park [157], reproduced from *Journal of Materials Science* with the permission of Chapman & Hall, publishers).

counts. It was suggested that a common rate-limiting step occurred in both EE and PIE [155, 157].

Fine mesh grids were placed in the region between the sample and the detector to allow a retarding potential energy analysis of the EE and PIE accompanying fracture. The derivative of the count-rate against retarding grid potential is the energy distribution of the emitted particles. Both charges appeared to have very similar energy distributions peaking near 0 eV, but with high energy particles ranging up to a few hundred electron volts. The similarity of the EE and PIE energy distributions again suggested a shared crucial mechanistic step. The presence of high energy particles suggested that charging of the fracture surface plays a role in the ejection of particles from the surface [155, 157].

The pulses for the ion and electron detectors, placed on opposite sides of an E-glass/epoxy sample, were tested for coincidence within $\frac{1}{2}$ and 100 μs time windows. No coincidence of any statistical significance was observed in either time frame [155, 157]. Measurements of time correlations of EE and PIE accompanying failure, with a $\frac{1}{2}$ μs window, were carried out on composite strands of E-glass/epoxy and Kevlar/epoxy. The ratio of true coincidence rate to positive ion count rate was 0.02 and 0.03 respectively. Unfilled and bead-filled rubbers had ratios between 0.04 and 0.22 [159].

The fracture of filaments [158] gave rapidly decaying time constants of the order of 10 μs for glass or graphite and 100 μs for Kevlar fibres. Bulk epoxy resin had a time constant of 25 μs. PIE time distributions for these materials were indistinguishable from the electron curves. In the fracture of fibre-reinforced epoxy strands the EE and PIE curves decayed for many seconds, with intense emitters giving detectable emission one hour after fracture. Simultaneous measurement of the PIE and EE from the same fracture event found that the decay curves were of similar shape but different intensities.

Kevlar fibres were tested in tension [162] and monitored for electrons, positive ions and photons. A stranded fracture surface was obtained with multiple FE peaks over several hundred microseconds. Fibres were also broken by stretching over a dull aluminium edge, which resulted in the identification of three forms of fracture. Firstly clean fracture, in which there was low total emission and rapid decay of the FE. Secondly transverse fracture accompanied by some splitting, resulting in greater EE and PIE counts and longer decay times. Thirdly extensive splitting and interfibrillar failure along with high total emission and long EE and PIE decay times. Total emission counts correlated well with the extent of damage as ascertained by microphotography.

During the testing of Kevlar and E-glass fibres the time-of-flight (TOF) of the leading edge (lightest) PIE was plotted against the inverse square root of the drift tube potential and the slope then used to calculate M/q (particle mass/particle charge). Values of 60 ± 20 amu were obtained for Kevlar and 48 ± 12 amu for E-glass. It was anticipated that a shoulder on the TOF distribution would reveal details of the heavier ion sizes emitted, but none was detected [164].

Tests were also undertaken [161, 164] to monitor both electron emission and acoustic emission (AE) during a single test on multiple (0°, $\pm 45°$ or 0°/90°/90°/0°) graphite–epoxy composites in flexure. There was a slow build-up of prefracture EE during microcracking of the resin and separation of tiny bundles of fibre at the tensile surface, a rapid rise in the emissions during fracture and a slow decay of emissions after fracture. Comparison of the EE and AE data revealed that a proportion of the EE data coincided with AE events, suggesting that EE was only detected from events at or near the surface and that AE was monitoring both internal and external fractures. EE is probably sensitive to interfacial fracture energy, as higher EE was detected in compounds with a tough rather than a brittle interphase. A low EE/AE ratio was identified in the 0°/90°/90°/0° composites due to the isolation of the failure from the vacuum by the 0° surface layers.

Dickinson *et al.* [163] have proposed a conceptual model for fractoemission based on simultaneous monitoring of electron, photon and radiowave emissions from either alumina-filled epoxy or filled (glass beads, single crystal sucrose or single crystal quartz) polybutadiene. The model involves the following steps:

(a) charge separation due to fracture;
(b) desorption of gases along the new crack surfaces;
(c) gas discharge in the crack;
(d) energetic bombardment of the new crack walls;
(e) thermally stimulated electron emission (TSEE) accompanied by electron stimulated desorption of ions and excited neutrals.

7.9. SUMMARY

Nuclear magnetic resonance and Fourier-transform infrared spectroscopy appear to offer tremendous potential for use as non-destructive testing techniques for fibre-reinforced plastics. Resonance Raman spectroscopy allied to polydiacetylene fibres, for use as an internal molecular strain gauge which can be monitored optically, should be invaluable for increasing the understanding of the mechanics of load transfer. The other techniques covered here have some potential as NDT methods but need to be more thoroughly researched before application in the field.

REFERENCES

1. L. R. Whittington, *Whittington's Dictionary of Plastics*, Second edition, SPE/Technomic Publishing Co, Westport, CT, 1978.
2. S. P. Potts and J. Preston, A cryogenic nuclear magnetic resonance gyroscope, *Journal of Naval Science*, July 1980, **6**(3), 188–200.
3. R. Richards and K. J. Packer, NMR spectroscopy in solids, *Phil. Trans. Royal Society of London*, 18 March 1981, A229(1452), 475–686; Discussion meeting held 18–19 June 1980, London, proceedings published separately as ISBN-0-85403-160-X
4. R. R. Eckman, Hydrogen and deuterium NMR of solids by magic-angle spinning, PhD thesis, Lawrence Berkeley Laboratory, October 1982; Report No. LBL-14200; DE83-003634.
5. H. J. Hackelöer, O. Kanert, H. Tamler and J. Th. M. de Hosson, Dynamical in-situ nuclear-magnetic-resonance tensile apparatus, *Review of Scientific Instruments* (USA), March 1983, **54**(3), 341–345.
6. G. Epstein and I. Weinberg, The application of nuclear spin resonance techniques to nondestructive testing/inspection of reinforced plastics, Ford Motor Company Aeronutronic Division Publication No. U-2123, 3 May 1963.
7. A. P. Stepanov, S. N. Novikov and R. N. Pletnev, Nuclear magnetic resonance of ^{11}B in an aluminoborosilicate glass fibre, *Soviet Physics: Solid State*, January 1970, **11**(7), 1660–1661; translation of: *Fizika Tverdogo Tela*, July 1969, **11**(7), 2049–2052.
8. A. P. Stepanov and S. N. Novikov, Changes in the coordination number of boron during heat-treatment of aluminium–borosilicate glass fibre, *Inorganic Materials*, 1972, **8**(6), 980–983; translation of *Izvestiya Akademii Nauk SSR, Neorganicheskie Materialov*, June 1972, **8**(6), 1120–1125.
9. I. P. Guseva, S. N. Novikov and A. P. Stepanov, Effect of heat treatment on the strength of aluminium–boron–silicate glass fibres, DRIC-T-6936, May 1983, BR88328, AD A131 435, N83-34042; translation of: *Fizika i Khimiya Obrabotki Materialov*, 1973, (4), 110–117.
10. D. Stefan, Molecular motions in polyblends and composites based on bisphenol-A polycarbonate, PhD thesis, University of Toronto, Canada, 1972. (Available on microfilm direct from the National Library of Canada at Ottawa.)
11. D. Stefan and H. L. Williams, Molecular motions in bisphenol-A polycarbonates as measured by pulsed NMR techniques, II: Blends, block copolymers and composites, *Journal of Applied Polymer Science*, 1974, **18**(5), 1451–1476.

12. F. G. Fabulyak, Yu. S. Lipatov, V. S. Kuznetsova and Z. M. Demidenko, Processes of dielectric and NMR relaxation in the surface layers of acrylate–epoxy–styrene compounds, DRIC-T-6934, May 1983, BR 88059, AD A130 152, N83-32940; Translation of *Voprosy Khimii I Khimicheskoi Tekhnologii*, 1974, **32**, 3–13.
13. Yu. S. Lipatov and F. G. Fabulyak, Studying the molecular motion in surface layers of PMMA and polystyrene by impulse NMR method, *Vysokomolekulyarnye Soedineniya*, July 1968, **10**(7), 1605–1616.
14. Yu. S. Lipatov, F. G. Fabulyak and G. P. Ovchinnikova, Study of the molecular mobility of components in the system: solidifying resin–polycaproamide; translation of: *Fiz. Khim. Polim. Kompositskii* (Conference), Kiev, 1974; Naukova Dumka Press, Kiev, 1974, pp. 74–79.
15. Yu. S. Lipatov and L. M. Sergeeva, *Adsorption of polymers*, John Wiley and Sons, New York, 1974.
16. Yu. S. Lipatov and F. G. Fabulyak, Molecular mobility in surface layers of a methylmethacrylate–styrene copolymer, *Polymer Science USSR*, 1969, **11**(4), 800–809; translation of: *Vysokomolekulyarnye Soedineniya*, 1969, **A11**(4), 708–716.
17. D. H. Kaelble and P. J. Dynes, Methods for detecting moisture-degradation in graphite–epoxy composites, *Materials Evaluation*, April 1977, **35**(4), 103–108.
18. J. D. King, W. L. Rollwitz and G. A. Matzkanin, Magnetic resonance methods for NDE, *Proc. 12th Symposium on NDE*, San Antonio, April 1979, pp. 1–12.
19. G. A. Matzkanin, Investigation of the effect of moisture on the mechanical properties of organic matrix composite materials using nuclear magnetic resonance, SWRI-15-5607-807: USAAVRADCOM-TR-81-F-5, May 1981, AD A100 426.
20. G. A. Matzkanin, Determination of moisture in fibre reinforced composites using pulsed NMR, AMMRC-MS-82-3, May 1982, AD A116 733; *Proc. Critical Review of Techniques for the Characterisation of Composite Materials*, ONR, MIT, Cambridge, June 1981, pp. 181–197.
21. G. A. Matzkanin, Applications of nuclear magnetic resonance to the NDE of composites, *Proc. 14th Symposium on NDE*, San Antonio, April 1983, pp. 270–286.
22. D. Lawing, R. E. Fornes, R. D. Gilbert and J. D. Memory, Temperature dependence of broadline NMR spectra of water-soaked, epoxy–graphite composites, *Journal of Applied Physics* (USA), October 1981, **52**(10), 5906–5907.
23. R. T. Fuller, A broadline NMR investigation of water absorbed by cured epoxy resin–graphite fibre composites, Master of Science in Physics, North Carolina State University at Raleigh, 1982; BLDSC shelfmark 82/26238.
24. R. T. Fuller, R. E. Fornes, and J. D. Memory, NMR study of water absorbed by epoxy resin, *Journal of Applied Polymer Science*, 1979, **23**(6), 1871–1874.
25. R. T. Fuller, S. Sherrow, R. E. Fornes, and J. D. Memory, Hydrogen exchange between water and epoxy resin, *Journal of Applied Polymer Science*, 1979, **24**(5), 1383–1385.

26. N. K. Batra and T. P. Graham, Measurement of water in hygrothermally degraded fibre-reinforced-epoxy composites by continuous wave (cw) nuclear magnetic resonance (NMR), *British Journal of Nondestructive Testing*, January 1983, **25**(1), 21–23.
27. C-H. Chiang, N-I. Liu and J. L. Koenig, Magic-angle cross-polarisation carbon-13 NMR study of aminosilane coupling agents on silica surfaces, *Journal of Colloid and Interface Science*, March 1982, **86**(1), 26–34.
28. A. Cholli, A. M. Zaper and J. L. Koenig, Applications of solid-state magic-angle NMR spectroscopy to fibre reinforced composites, *American Chemical Society National Meeting*, March 1983, pp. 215–216.
29. A. M. Evdokimov, V. V. Moskalev, V. L. Talanov, V. P. Kolonistov, A. S. Andreev and K. E. Perepelkin, ^{1}H-NMR study of molecular mobility in the interaction of chemical fibres with polymeric binders, *Mechanics of Composite Materials*, 1981, **16**(6), 642–646; translation of: *Mekhanika Kompozitnykh Materialov*, November–December 1981, (**6**), 982–986.
30. I. Sugiya, S. Kobayashi, S. Iwayanagi and T. Shibata, A broadline NMR study of the molecular motions in aromatic polyamides: poly(para-phenylene terephthalamide) and poly(meta-phenylene isophthalamide), *Polymer Journal*, 1982, **14**(1), 43–50.
31. G. Hempel and H. Schneider, ^{13}C-NMR investigations of the structure in solid polymers, *Pure and Applied Chemistry*, March 1982, **54**(3), 635–646; *22nd Microsymposium on Macromolecules*, IUPAC, Prague, July 1981.
32. G. I. Goryainov and A. I. Koltsov, Investigation of molecular mobility in polyimide fibres of various chemical structures by NMR; translation of: *Vysokomolekulyarnye Soedineniya*, 1982, **B24**(12), 910–913.
33. W. B. Alston, Characterisation of PMR-15 polyimide resin composition in thermooxidatively exposed graphite fibre composites, NASA-TM-81565: AVRADCOM-TR-80-C-10, 1980, N82-28524; Prox. 12th National SAMPE Technical Conference, Seattle, October 1980, pp. 121–137.
34. R. W. Lauver, W. B. Alston and R. D. Vannucci, Stability of PMR-polyimide monomer solutions, *34th Annual Reinforced Plastics/Composites Conference*, SPI, New Orleans, Jan./Feb. 1979, Paper 23A.
35. M. Kowalska and A. Wirsén, Chemical analysis of epoxy prepregs—market survey and batch control, *25th National SAMPE Symposium*, San Diego, May 1980, pp. 389–402.
36. J. A. Happe, R. J. Morgan and C. M. Walkup, ^{1}H, ^{19}F and ^{11}B NMR characterisation of BF_3: amine catalysts used in the cure of C-fibre epoxy prepregs, UCRL-89627, December 1983, DE84 006 139; Proc. 29th National SAMPE Symposium, Reno, April 1984, pp. 921–933.
37. J. A. Happe, R. J. Morgan and C. M. Walkup, NMR characterisation of BF_3-amine catalysts used in the cure of carbon fibre/epoxy prepregs, *Composites Technology Review*, Summer 1984, **6**(2), 77–82.
38. L. Banks and B. Ellis, The glass transition temperatures of highly crosslinked networks: cured epoxy resins, *Polymer*, September 1982, **23**(10), 1466–1472.
39. I. M. Brown and A. C. Lind, Study of the glass transition in crosslinked polymers using nuclear and electron magnetic resonance, McDonnell Douglas Research Report MDC-Q0633, 6 December 1977, AD A048 981.

40. I. M. Brown, A. C. Lind and T. C. Sandreczki, Magnetic resonance studies of epoxy resins and polyurethanes, McDonnell Douglas Research Report, MDC-Q0673, 3 May 1979, AD A073 590.
41. I. M. Brown, A. C. Lind and T. C. Sandreczki, Magnetic resonance studies of epoxy resins, McDonnell Douglas Research Report MDC-Q0721, 7 December 1980, AD A099 225.
42. I. M. Brown, A. C. Lind and T. C. Sandreczki, Magnetic resonance determinations of structure and reaction kinetics of epoxy/amine systems, McDonnell Douglas Research Report MDC-Q0759, 31 December 1981, AD A116 542.
43. A. N. Garroway, W. B. Moniz and H. A. Resing, ^{13}C-NMR in cured epoxies: benefits of magic angle rotation, *122nd ACS Organic Coatings and Plastics Chemistry Division*, San Francisco, 1976, **36**(2), 133–138.
44. A. N. Garroway, W. B. Moniz and H. A. Resing, High resolution ^{13}C-NMR in cured epoxy polymers: rotating frame relaxation, *Faraday Symposia— Chemical Society*, 1978, **13**, 63–74.
45. A. N. Garroway, W. B. Moniz and H. A. Resing, Carbon-13 NMR in organic solids: the potential for polymer characterisation, *ACS Symposium Series*, 1979, **103**, 67–87, Paper 4.
46. A. C. Lind, NMR study of inhomogeneities in amine cured epoxies, *Polymer Preprints*, 1981, **22**(2), 333–334.
47. C. F. Poranski, W. B. Moniz, D. L. Birkle, J. T. Kopfle and S. A. Sojka, Carbon-13 and proton NMR spectra for characterizing thermosetting polymer systems, *NRL Report 8092*, 20 June 1977, AD A044 214.
48. H. A. Resing and W. B. Moniz, A first step toward high resolution ^{13}C-NMR spectroscopy of intractable polymers: epoxies, *Macromolecules*, July–August 1975, **8**(4), 560–561.
49. H. A. Resing, A. N. Garroway, D. C. Weber, J. Ferraris and D. Slotfeldt-Ellingsen, 13C-NMR of solid polymers and of solids related to polymer composites, *Pure and Applied Chemistry*, March 1982, **54**(3), 595–610; *22nd Microsymposium on Macromolecules*, IUPAC, Prague, July 1981.
50. J. Schaeffer, E. O. Stejskal, M. D. Sefcik and R. A. McKay, Applications of high resolution ^{13}C and ^{15}N n.m.r. in solids, *Phil. Trans. Royal Society of London*, 18 March 1981, **A299**(1452), 593–608; discussion meeting held 18–19 June 1980, London; Proceedings published separately as ISBN-0-85403-160-X, pp. 117–132.
51. V. P. Tarasov, Yu. N. Smirnov, L. N. Yerofeyev, V. I. Irzhak and B. A. Rozenberg, On the nature of segmental mobility of dense-network epoxy polymers, *Polymer Science USSR*, 1982, **24**(11), 2732–2736; translation of: *Vysokomolekulyarnye Soedineniya*, 1982, **A24**(11), 2379–2382.
52. H. L. Tighzert, P. Berticat, B. Chabert and Q. T. Pham, Etude par RMN ^{1}H à 350 MHz de la polycondensation en masse entre les résines époxydes et des amines aromatiques, *Comptes Rendus au 3ème Journées Nationales sur les Composites* (JNC3), Paris, September 1982, 375–383.
53. R. V. Damadian, Tumour detection by nuclear magnetic resonance, *Science*, 19 March 1971, **171**(3976), 1151–1153.
54. P. C. Lauterbur, Image formation by induced local interactions: examples

employing nuclear magnetic resonance, *Nature*, 16 March 1973, **242**(5394), 190–191.
55. P. Mansfield and A. A. Maudsley, Medical imaging by NMR, *British Journal of Radiology*, March 1977, **50**(591), 188–194.
56. W. S. Hinshaw, P. A. Bottomley and G. N. Holland, Radiographic thin-section image of the human wrist by nuclear magnetic resonance, *Nature*, December 1977, **270**(5639), 722–723.
57. R. Herman, A chemical clue to disease, *New Scientist*, 15 March 1979, **81**(1146), 874–877.
58. R. J. P. Williams, E. R. Andrews and G. K. Radda (editors), Nuclear magnetic resonance of intact biological systems, *Phil. Trans. Royal Society of London*, 25 June 1980, **B289**(1037), 379–559; discussion meeting held 14–15 March 1979, London.
59. L. Kaufman, L. E. Crooks and A. Margulis, *NMR imaging in medicine*, Igaku-Shoin Medical Publishing Inc., New York, 1981.
60. P. Mansfield and P. G. Morris, NMR imaging in biomedicine: supplement 2 to *Advances in Magnetic Resonance*, Academic Press, New York and London, 1982.
61. Z. H. Cho, H. S. Kim, H. B. Song and J. Cumming, Fourier transform nuclear magnetic resonance tomographic imaging, *Proceedings of IEEE*, October 1982, **70**(10), 1152–1173.
62. J. Jaklovsky and W. H. Oldendorf, *NMR imaging—a comprehensive bibliography*, Addison-Wesley, Reading, MA, January 1984.
63. W. P. Rothwell, D. R. Holecek and J. A. Kershaw, NMR imaging: study of fluid absorption by polymer composites, *Journal of Polymer Science: Polymer Letters Edition*, May 1984, **22**(5), 241–247.
64. A. Abragam, *The principles of nuclear magnetism*, Clarendon Press, Oxford, 1961.
65. R. R. Hewitt and B. Mazelsky, Nondestructive inspection of reinforced plastic structures by the nuclear quadrupole resonance method, Aerospace Research Associates Report #77, 1 November 1966, AD645 562.
66. R. R. Hewitt and B. Mazelsky, Nuclear quadrupole resonance as a nondestructive probe in polymers, *Journal of Applied Physics*, August 1972, **43**(8), 3386–3392.
67. J. M. Hays, *Reactive free radicals*, Academic Press, London, 1974.
68. I. N. Levine, *Molecular spectroscopy*, Wiley-Interscience, New York, 1975, pp. 366–367.
69. D. Robson, F. Y. I. Assabghy, D. J. E. Ingram and P. G. Rose, Determination of carbon fibre structure by electron spin resonance, *Nature*, 4 January 1969, **221**(5175), 51–52.
70. D. Robson, F. Y. I. Assabghy, D. J. E. Ingram and P. G. Rose, Electron spin resonance studies of carbon fibres based on polyacrylonitrile, *Proc. 3rd London Conf. Industrial Carbons and Graphite*, April 1970, 453–455.
71. D. Robson, F. Y. I. Assabghy and D. J. E. Ingram, An electron spin resonance study of carbon fibres based on polyacrylonitrile, *Journal of Physics D: Applied Physics*, September 1971, **4**(9), 1426–1438.
72. D. Robson, F. Y. I. Assabghy and D. J. E. Ingram, Some electronic prop-

erties of polyacrylonitrile-based carbon fibres, *Journal of Physics D: Applied Physics*, January 1972, **5**(1), 169–179.
73. D. Robson, F. Y. I. Assabghy, E. G. Cooper and D. J. E. Ingram, Electronic properties of high-temperature carbon fibres and their correlations, *Journal of Physics D: Applied Physics*, October 1973, **6**(10), 1822–1834.
74. P. A. Pschenichkin, V. S. Samoilov, V. A. Mikhailova and S. G. Zaichikov, Investigation of the structure of carbonised polyacrylonitrile (PAN) fibres by the EPR method; translation of: *Trudy Vsesoyuznogo Naucho-issledovatels'kogo Proekt-Tekhnologi Institut Elektrougol Izdelii*, 1972, (2), 131–135.
75. A. S. Fialkov, P. A. Pschenichkin, N. A. Melnikova, S. G. Zaichikov, V. S. Samoilov and T. N. Zhuikova, Paramagnetic electron resonance in carbon fibre, *Fibre Chemistry*, 1974, (2), 160–162; *Khimicheskie Volokna*, March–April 1974, (2), 31–33.
76. A. S. Kotosonov, V. S. Tverskoi and V. I. Frolov, Electron spin resonance of polyacrylonitrile-based carbon fibres, *4th Int. London Conf. Carbon and Graphite*, September 1974; proceedings published 1976, pp. 508–509.
77. A. A. Bright and L. S. Singer, Electronic and structural characteristics of carbon fibres from mesophase pitch, *13th Biennial Carbon Conference*, Irvine, July 1977, 100–101.
78. A. A. Bright and L. S. Singer, The electronic and structural characteristics of carbon fibres from mesophase pitch, *Carbon*, 1979, **17**(1), 59–69.
79. J. W. McClure, S. D. Elegba and B. Hickman, Theory of the diamagnetism and g-shift of carbon fibres, *15th Biennial Carbon Conference*, Philadelphia, June 1981, pp. 10–11; extended abstracts programme.
80. J. B. Jones and L. S. Singer, Electron spin resonance and the structure of carbon fibres, *Carbon*, 1982, **20**(5), 379–385. *15th Biennial Carbon Conference*, Philadelphia, June 1981, 306–307.
81. J. R. Brown and D. K. C. Hodgeman, An e.s.r. study of the thermal degradation of Kevlar 49 aramid, *Polymer*, March 1982, **23**(3), 365–368.
82. I. M. Brown and T. C. Sandreczki, Characterisation of Kevlar 49 fibres by electron paramagnetic resonance, McDonnell Douglas Research Report MDC-Q0775, 20 June 1982; UCRL-15512, DE83 004 218.
83. I. M. Brown, T. C. Sandreczki and R. J. Morgan, Electron paramagnetic resonance studies of Kevlar 49 fibres: stress-induced free radicals, *Polymer*, June 1984, **25**(6), 759–765.
84. Y. Ogo, Y. Okuri and Y. Miura, Mechanochemical reactions of FRP at high pressures; translation of: *Zairyo*, July 1983, **32**(358), 808–812.
85. Y. Ogo, Y. Okuri and Y. Nakai, Mechanochemical reactions of polymethylmethacrylate and polystyrene at high pressure, DRIC-T- 7307, 1984; BR 92544; translation of: *Zairyo*, 1981, **30**(336), 893–897.
86. E. A. Barbashov, V. A. Bogatov, R. I. Kozyreva, B. A. Kiselev, B. I. Panshin, B. V. Perov and M. A. Chubarov, Electron paramagnetic resonance and infrared spectroscopy studies of the relations between changes in the mechanical properties and the chemical structure of epoxy-bonded reinforced plastics under the influence of electron bombardment, *Soviet Materials Science*, 1969, **5**(3), 245–248; *Fiziko-Khimicheskaya Mekhanika Materialov*, 1969, **5**(3), 318–322.
87. S. G. Klimanov and S. E. Rubakova, An investigation of the anisotropy of glass fibres; translation of: *Fizika i Khimiya Stekla*, 1981, **7**(2), 252–255.

88. H. Kato, A. Ito, M. Tanaka and T. Miyahara, Changes in the optic glass fibre due to x-ray irradiation-electron spin resonance study, *Japanese Journal of Tuberculosis and Chest Diseases*, 1976, **20**(1–2), 25–37.
89. L Cociu, A. Marton and A. Nicula, Study of the ageing of stratified electroinsulating materials by the electron spin resonance method, *Studia Univ. Babes-Bolyai, Physica*, 1980, **25**(1), 26–28.
90. K. R. Schaffer, Characterisation of a cured epoxy resin exposed to high energy radiation with electron spin resonance, MS Thesis, North Carolina State University at Raleigh, 1981; BLDSC shelfmark 84/15441.
91. (a) McGraw-Hill Encyclopaedia of Science and Technology, 5th Edition, 1982, Volume 11, pages 387–391; (b) I. Suzuki and M. Tasumi, Infrared and Raman spectroscopy literature database—a bibliography covering the period 1 Jan 82–30 Sep 84, *Journal of Molecular Structure*, December 1985, **132**, 1–476; (c) I. Suzuki and M. Tasumi, Infrared and Raman spectroscopy literature database—a bibliography covering the period 1 Oct 84–31 May 86, *Journal of Molecular Structure*, December 1986, **155**, 1–322.
92. J. Behringer, Theories of resonance Raman scattering, *Molecular Spectroscopy*, 1974, **2**, 100–172.
93. J. Behringer, Experimental resonance Raman spectroscopy, *Molecular Spectroscopy*, 1975, **3**, 163–280.
94. F. Tuinistra and J. L. Koenig, Characterisation of graphite fibres surfaces with Raman spectroscopy, *Journal of Composite Materials*, October 1970, **4**(4), 492–499.
95. J. L. Koenig and P. T. K. Shih, Raman studies of the glass fibre–silane–resin interface, *Journal of Colloid and Interface Science*, June 1971, **36**(2), 247–253.
96. P. T. K. Shih, Raman studies of glass reinforced polyester composites, PhD thesis, Case Western Reserve University, 1973, DA7316159.
97. J. L. Koenig, P. T. K. Shih and P. Lagally, Raman and infrared spectroscopic studies of the reactions of silica and glass surfaces, *Materials Science and Engineering*, August 1975, **20**(2), 127–135.
98. H. Ishida, J. L. Koenig, B. Asumoto and M. E. Kenney, Application of UV-resonance Raman spectroscopy to the detection of monolayers of silane coupling agent on glass surfaces, *Polymer Composites*, April 1981, **2**(2), 75–80.
99. C. Galiotis, R. J. Young and D. N. Batchelder, A resonance Raman spectroscopic study of the strength of the bonding between an epoxy resin and a polydiacetylene fibre, *Journal of Materials Science Letters*, June 1983, **2**(6), 263–266.
100. C. Galiotis, R. J. Young and D. N. Batchelder, Strain measurement in polymer crystals using resonance Raman spectroscopy, *Biennial Conference on Physical Aspects of Polymer Science*, IoP (PPG) and RSC, Reading, 14–16 September 1983, abstract only.
101. C. Galiotis, R. J. Young, P. H. J. Yeung and D. N. Batchelder, A study of model polydiacetylene/epoxy composites. Part 1: the axial strain in the fibre, *Journal of Materials Science*, November 1984, **19**(11), 3640–3648.
102. R. J. Young, C. Galiotis and P. H. J. Yeung, Development of high-modulus polydiacetylene fibres for use in fibre-reinforced composites, AD A140 912, February 1984.

102a. R. J. Young, C. Galiotis and D. N. Batchelder, Use of Raman spectroscopy to follow the micromechanics of deformation in polymer fibres and composites, Meeting on *The physics of interfaces in reinforced thermoplastics*, Inst. Phys., London, 15 May 1986, Paper 2, abstract only.
103. Y. T. Liao and J. L. Koenig, Applications of Fourier transform infrared spectroscopy to the study of fibre composites, Chapter 2 of: G. Pritchard (editor), *Developments in Reinforced Plastics—4*, Elsevier Applied Science Publishers, London and New York, 1984, pp. 31–87.
104. J. W. Cooley and J. W. Tukey, An algorithm for machine calculation of complex Fourier series, *Mathematics of Computation*, April 1965, **19**(90), 297–301.
105. E. Oran Brigham, *The Fast Fourier Transform*, Prentice-Hall Inc., Englewood Cliffs, NJ, 1974.
106. H. Lee and K. Neville, *Handbook of Epoxy Resins*, McGraw-Hill Book Co., New York, 1967, p. 13–1.
107. M. K. Antoon, Fourier transform infrared investigation of the structure and moisture stability of the epoxy matrix in glass-reinforced composites, PhD thesis, Case Western Reserve University, January 1980, DA8013869.
108. S. C. Lin, B. J. Bulkin and E. M. Pearce, Epoxy resins III: application of FTIR to degradation studies of epoxy systems, *Journal of Polymer Science: Polymer Chemistry Edition*, 1979, **17**(10), 3121–3148.
109. L. Buckley and D. Roylance, Kinetics of a sterically hindered amine-cured epoxy resin system, *Polymer Engineering and Science*, 1982, **22**(3), 154–159.
110. J. F. Sprouse, B. M. Halpin and R. E. Sacher, Cure analysis of epoxy composites using FT–IR, AMMRC-TR-78-45, November 1978, AD A067 197.
111. M. K. Antoon, K. M. Starkey and J. L. Koenig, Applications of FT-IR spectroscopy to quality control of the epoxy matrix, *ASTM Special Technical Publication 674*, June 1979, pp. 541–552; *5th Conference on Composite Materials: Testing and Design*, New Orleans, March 1978.
112. R. H. Pater and D. A. Scola, Nondestructive analysis of composite surface contamination by Fourier transform infrared spectroscopy, *11th National SAMPE Technical Conference*, Boston, MA, November 1979, pp. 151–165.
113. B. E. Walker, C. T. Ewing and R. R. Miller, Nondestructive testing for void content in glass-filament-wound composites, US Naval Research Laboratory Report NRL-6775, 4 October 1968, AD 679 573.
114. G. A. George, R. E. Sacher and J. F. Sprouse, Photooxidation and photoprotection of the surface resin of a glass fibre–epoxy composite, *Journal of Applied Polymer Science*, August 1977, **21**(8), 2241–2251.
115. P. R. Young, B. A. Stein and A. C. Chang, Resin characterisation in cured graphite fibre reinforced composites using diffuse reflectance FTIR, *28th National SAMPE Symposium*, Anaheim, April 1983, Paper 101, pp. 824–837.
116. H. Ishida and J. L. Koenig, The reinforcement mechanism of fibre-glass reinforced plastics under wet conditions: a review, *Polymer Engineering and Science*, mid-February 1978, **18**(2), 128–145.

117. V. R. Regel', A. D. Gabaraeva, N. N. Filippov and A. M. Leksovskii, Measuring the stresses in fibres of a loaded composite material by the method of infrared spectroscopy, *Polymer Mechanics*, September–October 1977, **13**(5), 696–701; *Mekhanika Polimerov*, September–October 1977, **13**(5), 832–837; Presented at the Third All-Union Conference on Polymer Mechanics, Riga, 1976.
118. M. Ishida and J. L. Koenig, FTIR spectroscopic study of the silane coupling agent/porous silica interface, *Journal of Colloid and Interface Science*, May 1978, **64**(3), 555–564.
119. H. Ishida and J. L. Koenig, FTIR spectroscopic study of the structure of silane coupling agent on E-glass fibre, *Journal of Colloid and Interface Science*, May 1978, **64**(3), 565–576.
120. H. Ishida and J. L. Koenig, An investigation of the coupling agent/matrix interface of fibreglass reinforced plastics by FTIR, *Journal of Polymer Science: Polymer Physics Edition*, April 1979, **17**(4), 615–626.
121. H. Ishida and J. L. Koenig, Molecular organisation of the coupling agent interphase of fibre-glass reinforced plastics, *Journal of Polymer Science: Polymer Physics Edition*, October 1979, **17**(10), 1807–1813.
122. C-H. Chiang and J. L. Koenig, Chemical reactions occurring at the interface of epoxy resin and aminosilane coupling agents in fibre reinforced composites, *35th Annual SPI Reinforced Plastics/Composites Institute Conference*, New Orleans, February 1980, Paper 23-D.
123. H. Ishida and J. L. Koenig, Effect of hydrolysis and drying on the siloxane bonds of a silane coupling agent deposited on E-glass fibres, *Journal of Polymer Science: Polymer Physics Edition*, February 1980, **18**(2), 233–237.
124. C-H. Chiang, H. Ishida and J. L. Koenig, The structure of γ-aminopropyltriethoxysilane on glass surfaces, *Journal of Colloid and Interface Science*, April 1980, **74**(2), 396–404.
125. H. Ishida and J. L. Koenig, An FTIR spectroscopic study of the hydrolytic stability of silane coupling agents on E-glass fibres, *Journal of Polymer Science: Polymer Physics Edition*, September 1980, **18**(9), 1931–1943.
126. H. Ishida and J. L. Koenig, New spectroscopic techniques for studying glass surfaces; in: D. E. Leyden and W. Collins (editors), *Midland Macro Molecular Monographs No. 7—Silylated surfaces*, Gordon and Breach, New York, August 1980, pp. 73–98.
127. C. Chiang and J. L. Koenig, Comparison of primary and secondary aminosilane coupling agents in anhydride-cured epoxy fibreglass composites, *36th Annual SPI Reinforced Plastics/Composites Institute Conference*, Washington, DC, February 1981, Paper 2-D.
128. J. L. Koenig and C-H. Chiang, In-situ analysis of the interface; in: J. C. Seferis and L. Nicolais (editors), *The role of the polymeric matrix in the processing and structural properties of composite materials*, Plenum, New York and London, 1983, pp. 503–516; *Proceedings of the Joint US–Italy Symposium on Composite Materials*, Capri, Italy, 15–19 June 1981.
129. Chwan-Hwa Peter Chiang, FTIR spectroscopic investigation of the fibre–matrix interface in fibreglass reinforced epoxy composites, PhD thesis, Case Western Reserve University, January 1981; DA8109576, N81-28174.
130. S. Naviroj, J. Koenig and H. Ishida, The formation of linear-chain amine–

bicarbonate salt in a partially cured aminosilane and its influence on the reactivity with an epoxy resin, *37th Annual SPI Reinforced Plastics/Composites Institute Conference*, Washington, DC, January 1982, Paper 2-C.
131. H. Ishida, S. Naviroj, S. K. Tripathy, J. J. Fitzgerald and J. L. Koenig, The structure of an aminosilane coupling agent in aqueous solutions and partially cured solids, *Journal of Polymer Science: Polymer Physics Edition*, 1982, **20**(4), 701–718, AD A100 756.
132. C-H. Chiang and J. L. Koenig, Spectroscopic characterisation of the matrix–silane coupling agent interface in fibre reinforced composites, *Journal of Polymer Science: Polymer Physics Edition*, November 1982, **20**(11), 2135–2143.
133. S. R. Culler, H. Ishida and J. L. Koenig, The use of infrared methods to study polymer interfaces, CWRU/DMS/TR-7, 9 December 1982, AD A123 021.
134. H. Ishida, S. Naviroj and J. L. Koenig, The influence of a substrate on the surface characteristics of silane layers; in: K. L. Mittal (editor), *Physicochemical aspects of polymer surfaces*, Plenum, New York and London, 1983, vol. 1, pp. 91–104; Proceedings of an International Symposium, ACS, New York, August 1981.
135. R. T. Graf, J. L. Koenig and H. Ishida, The influence of interfacial structure on the flexural strength of E-glass reinforced polyester, *Journal of Adhesion*, 1983, **16**(2), 97–113.
136. S. R. Culler, H. Ishida and J. L. Koenig, Nondestructive FTIR sampling technique to study glass fibre composite interfaces, CWRU/DMS/TR-9, 20 June 1983; AD A129 871.
137. S. R. Culler, S. Naviroj, H. Ishida and J. L. Koenig, Analytical and spectroscopic investigation of the interaction of carbon dioxide with amine functional silane coupling agents on glass fibres, *Journal of Colloid and Interface Science*, 1983, **96**(1), 69–79, CWRU/DMS/TR-8, 31 December 1982; AD A124, 151.
138. S. R. Culler, H. Ishida and J. L. Koenig, Nondestructive FTIR sampling technique to study glass fibre composite interfaces, *Applied Spectroscopy*, January/February 1984, **38**(1), 1–7.
139. S. Naviroj, S. R. Culler, J. L. Koenig and H. Ishida, Structure and adsorption characteristics of silane coupling agents on silica and E-glass fibre: dependence on pH, *Journal of Colloid and Interface Science*, 1984, **97**(2), 308–317.
140. S. R. Culler, H. Ishida and J. L. Koenig, Silane interphase of composites: effects of process conditions on gamma-aminopropyltriethoxysilane, CWRU/DMS/TR-16, 4 September 1984; AD A145 578, N85–11143.
141. M. G. Kamenskii, V. A. Golubev, V. P. Korkhov, A. A. Kul'kov, Yu. M. Molchanov and E. F. Kharchenko, Study of the structure of organic plastics reinforced with polyheteroarylene fibres, *Mechanics of Composite Materials*, January–February 1983, **19**(1), 51–54. *Mekhanika Kompozitnykh Materialov*, January–February 1983, (**1**), 61–65.
142. M. F. Grundon and H. B. Henbest, *Organic chemistry—an introduction*, fifth impression, Macdonald Technical and Scientific, London, 1971.
143. Wu Xu-Quig, Gan Shu-Bang and Zhang Yuan-Min, The reinforcement

mechanism of the interface of glass fibre reinforced plastics, *14th Reinforced Plastics Congress*, BPF, Brighton, November 1984, Paper 43, pp. 199–201.
144. D. M. Considine (editor), *Van Nostrand's Scientific Encyclopaedia*, sixth edition, Van Nostrand Reinhold Company, New York, 1983.
145. G. A. George and D. M. Pinkerton, Studies of the characterisation of organic materials by non-stationary chemiluminescence techniques, AMMRC-MS-82-3, May 1982; AD A116 733; *Proc. Critical Review: Techniques for the characterisation of composite materials*, ONR, Massachussetts, June 1981, Session III, pp. 291–309.
146. C. J. Wolf, D. L. Fanter and M. A. Grayson, Ageing of polymers and composites, McDonnell Douglas Research Report MDC-Q0743, 21 July 1981; AD A111 625.
147. C. J. Wolf, M. A. Grayson and D. L. Fanter, Ageing of polymeric materials, McDonnell Douglas Research Report MDC-Q0798, 31 December 1982; AD A132 250.
148. R. B. Rickards, G. A. Teters, and Z. T. Upitis, Models of the failure of composites having various reinforcement structures, *Mechanics of Composite Materials*, March–April 1979, (2), 162–167; *Mekhanika Kompozitnykh Materialov*, March–April 1979, (2), 222–227.
149. R. B. Rickards, G. A. Teters and Z. T. Upitis, Models of the failure of composites having various reinforcement structures, in: G. C. Sih and V. P. Tamuzs (editors), *Fracture of Composite Materials*, Sijthoff and Noordhoff, Aalphen aan den Rijn, 1979 (Russian version published by Zinatne, Riga); *Proc. 1st joint USSR–USA Symposium*, Riga, September 1978.
150. Z. T. Upitis, U. E. Krauya and Y. A. Lanson, The mechanoluminescence of fibreglass tubes in the plane stressed state, DRIC-T-7014, August 1983. BR 89832; AD A135 680; N84-18673; translation of *Mekhanika Kompozitnykh Materialov*, 1980, (3), 552–556.
151. U. E. Krauya, Z. T. Upitis, R. B. Rickards, G. A. Teters and Ya. L. Yansons, Monitoring destruction of fibreglass plastics with mechanoluminescence, *Mechanics of Composite Materials*, March–April 1981, (2), 236–241; *Mekhanika Kompozitnykh Materialov*, March–April 1981, (2), 325–331.
152. G. A. Teters, U. E. Krauja, R. B. Rickards and Z. T. Upitis, Mechanical luminescence study of composite fracture in a plane stressed state; in: G. C. Sih and V. P. Tamuzs (editors), *Fracture of Composite Materials*, Martinus Nijhoff, The Hague, 1982; *Proceedings of the 2nd Joint USA–USSR Symposium*, Bethlehem, PA, March 1981.
153. G. A. Teters, U. E. Krauja, R. B. Rickards and Z. T. Upitis, Mechanical luminescence study of composite fracture in a plane stressed state, *Mechanics of Composite Materials*, November 1982, **18**(3), 379–387; *Mekhanika Kompozitnykh Materialov*, May–June 1982, (3), 537–545.
154. G. Ya. Menzheres, A. N. D'Yakova, M. S. Kligshtein, E. G. Moisya and F. G. Fabulyak, Investigation by the molecular probe method of the changes in the packing density of filled epoxide compositions during hardening, *Theoretical and Experimental Chemistry* (USA), 1982, **18**(2), 219–222; *Teoreticheskaya i Eksperimental'naya Khimiya*, March–April 1982, **18**(2), 253–256.

155. J. T. Dickinson, E. E. Donaldson and M. K. Park, The emission of electrons and positive ions from fracture of materials, AD A097 390, April, 1981.
156. J. T. Dickinson, Fracto-emission from composites, AMMRC-MS-82-3, May 1982; AD A116 733; *Proc. Critical Review: Techniques for the characterisation of composite materials*, ONR, Massachussetts, June 1981, Session IV, pp. 371–386.
157. J. T. Dickinson, E. E. Donaldson and M. K. Park, The emission of electrons and positive ions from fracture of materials, *Journal of Materials Science*, October 1981, **16**(10), 2897–2908.
158. J. T. Dickinson, M. K. Park, E. E. Donaldson and L. C. Jensen, Fracto-emission accompanying adhesive failure, *Journal of Vacuum Science and Technology*, March 1982, **20**(3), 436–439.
159. J. T. Dickinson, L. C. Jensen and M. K. Park, Time correlations of electron and positive ion emissions accompanying and following fracture of a filled elastomer, *Applied Physics Letters*, 1 September 1982, **41**(5), 443–445.
160. J. T. Dickinson, Fracto-emission from graphite–epoxy composites, NASA-CR-173 845, July 1983; N84-31291 (compilation of references 161–164).
161. J. T. Dickinson, A. Jahan-Latibari and L. C. Jensen, Electron emission and acoustic emission from the fracture of graphite/epoxy composites, *Journal of Materials Science*, January 1985, **20**(1), 229–236.
162. J. T. Dickinson, A. Jahan-Latibari and L. C. Jensen, Fracto-emission from single fibres of Kevlar, *Journal of Materials Science*, May 1985, **20**(5), 1835–1841.
163. J. T. Dickinson, L. C. Jensen and A. Jahan-Latibari, Fracto-emission: the role of charge separation, *Journal of Vacuum Science and Technology*, April–June 1984, **A2**(2), 1112–1116.
164. J. T. Dickinson, A. Jahan-Latibari and L. C. Jensen, Fracto-emission from fibre-reinforced and particulate-filled composites; in: H. Ishida and G. Kumar (editors), Molecular characterization of composite interfaces, *Polymer Science and Technology*, vol. 27, Plenum Press, New York & London, 1985, pp. 111–131; *Proc. Symposium: Polymer composites: interfaces*, American Chemical Society, Seattle, March 1983.

Index

Acoustic
 coupling, 36
 damping, 50
 emission
  advantages of, 47–8
  attenuation in, 35
  brittle materials, 28–9
  CFRP, 52–5
  component emission, 55
  composites, from, 48–58
  concrete, 30, 32
  continuous monitoring, 47
  data recording, 44–6
  degradation mechanisms in, 30–4
  distributions, of amplitudes, 53–4
  ductile materials, 28–9
  electric emission, 258
  energy in, 33–4
  Felicity effect, 30–1
  frequency analysis, 54
  frequency, for operation, 36
  GRP, and, 52–5
  Kaiser effect, 30–1
  load plots, 28–9
  materials evaluation, 46, 50–5
  mode conversion, 34–5
  periodic inspection, 47
  principles, 25–8
  proof testing, 46–7
  propagation, 34–5
  quality assurance, 46, 58
  sensors, 27, 35–6, 38

Acoustic—*contd.*
 emission—*contd.*
  signal processing, 38, 40–4
   analogue, 38, 40, 45
   digital, 40, 42, 44, 45–6
  source characterisation, 53–5
  stress intensity, 48
  stress wave factor, 48
  structural monitoring, 55–6
  spectral method, and NDT, 195
Acoustical holography, 136–7
 composites and, 137
 principles of, 136–7
Active heating, in thermal NDT
  methods, 65–6
 advantages of, 66
AE. *See* Acoustic emission
Aerofoil sections, 88–9
Alignment, fibre, 11–12
Aluminium alloys, 28
Amplitude
 analysis, in AE, 42–3
 distribution, in AE, 53–4
Analogue AE, 30, 40, 45
 burst activity, 40
 continuous activity, 40
Angle of diffraction. *See* Bragg's
  angle of diffraction
Aramid-fibre composite, 20
Arbitrary units of light emission, See
  AULE
Atomic orbitals, 208–9

AULE, 251–3
Automated fringe analysis, 125–9
Autoradiography, 5

Band-selectable Fourier analysis. *See*
  ZOOM analysis
Bar-and-spacing grating, 119
Barium lead sulphate, 6
Beams, cut from laminated plates,
  167–9
Bicycle pedal crank, 99
BO_3, 211–13
Bonded composite structures, 118
Boron fibres, 11–12
Bottles, glass, 46
Bragg's angle of diffraction, 8–9
Brittle lacquer, 144–5
  composites and, 145
Brittleness, and AE, 28–9
Broadband sensors, 36
'Bucket trucks', 56
Burst activity, in AE, 40–1

Calcium tungstate, 6
Californium-252, 5, 18
Carbon fibres, 25, 98–100
  epoxy bond, thermography, 86–7
  ESR of, 226–9
Carboxy/epoxy composites, 11–14, 20
  moisture determination, 18–20
CARP code, 30, 47, 49, 51, 56
Celion-6000, 221
Ceramics, 28
  'dunting', 26
CFRP, 20–1, 88–9, 91–3
  plates, and vibration, 180–2
  tubes, natural frequency, 177–9
Chemical imaging. *See* Liquid crystals
Chemical spectroscopy, 207–70
  ESR, 225–35
  FE, 255–8
  FTIR, 239–48
  luminescence, 249–54
  NMR, 208–25
  Raman, 235–9
  UV, 248–9

Chemiluminescence, 250–1
CL. *See* Chemiluminescence
Classical holography, 130–6
  composites, and, 135
  principles of, 130–2
  techniques of, 134–5
'Clinking', castings. *See* Martensitic
  transformations
Cobalt-60, 5
Coin-tap test, 187–9
Committee on Acoustic emission
  monitoring of Reinforced
  Plastics. *See* CARP code
Composite materials, vibration
  properties, 153–60
  unidirectional, 154–60
Composites. *See* Fibre-reinforced
  plastics composites *and also*
  specific names of
Computer-aided tomography, 17,
  222–3
Computers, 113
  AE, and, 45–7
  assisted radiography, 10
  software. *See* Software, computer
  thermal NDT modelling, 69–79
Concrete, and AE, 30, 32
Continuous structures, and vibration,
  163–71
  beams, from laminated plates, 167–9
  laminated plates, 169–71
  vibration types, 163–7
Copper coatings, 94
Corona discharge, 201–6
Cracks
  detection, by AE, 30–1
  vibration, and, 171–5
Cross polarisation (CP) spectroscopy.
  See Proton-enhanced nuclear
  induction spectroscopy
Crurix RPI, 13

Damping, 182
  cracks and, 171–5
  measurement methods, 176–8
  modulus, 171–5
  specific damping capacity, 153–4

Defect detection
  opaque penetration, 14–16
  thermal NDT, 71–2
Deformation
  contours, and moiré, 121
  out-of-plane, 113–17, 139–41
  shape, 113–17, 139–41
Delamination, 12–16
  corona discharge, and, 201–3
  infra-red camera, and, 87
Depth measurements, by
    triangulation, 110–12
Deuterium double quantum
    decoupling, 211
DIB. See Diiodobutane
Diffraction. See X-rays, diffraction
Diffractometers, 8–9
Digital AE, 40, 42, 44–6
  amplitude analysis, 42
  energy analysis, 42
  event counting, 40
  event duration, 42
  ringdown counting, 40, 42
  rise time, 42
  source location, 44
Diiodobutane, 12–13, 15
2DT program, 79–81, 83–4
Ductility, and AE, 28–9
'Dunting', ceramics. See Thermal
    shock

EATF, 67
Edge temperature gradient, 78
EE. See Electron emission
Electromagnetic interference (EMI),
    36
Electromagnetic radiation, 3–4
Electron emission (EE), 254–8
  acoustic emission, and, 258
Electron energy, 2–4
Electron spin resonance, 225–
    35
  carbon fibres, of, 226–9
  glass fibres, of, 233–5
  Kevlar fibres, of, 229–33
  resin, epoxy, of, 235
Electron volt, 4–5

Energy
  acoustic, 33–4, 42
  heat. See Thermal NDT methods
Enlargement, degree of, 6–8
Epoxy resin, 11
  composite degradation, and FTIR,
    242–5
  ESR of, 235
  FTIR, in cure, 240–2
ESR. See Electron spin resonance
Event counting, in AE, 40
Excitation (vibration)
  each test point, at, 186–96
    acoustic spectra, 195
    coin-tap test, 187–9
    impedance, 189–93
    membrane resonance, 193–5
    velocimetry, 195
  single points, at, 183–6
    local amplitude, 185–6
    vibrothermography, 183–5
Exposure, in radiography, 8
Externally applied thermal fields. See
    EATF

FE. See Fracto-emission
Felicity effect, 30–1, 51, 57
  definition, 51, 57
Fiberite T300/1034E, 13
Fibre
  alignment, 11–12, 65
  flaws, 11–12
  matrix interface, FTIR, 245–8
  optics, 107
  orientation, by thermal NDT,
    89–101
  stress, and FTIR, 245
  volume fractions, 11, 21
Fibre-reinforced plastic composites
  acoustic emission of, 25–63
  defect detection, X-rays, 14–16
  delamination, 13–16
  fibre content determination, 20–1
  fibre flaws, 11–12
  fibre volume fraction, 11
  moisture determination, 18–20
  optical methodology, 105–49

Fibre-reinforced plastic
    composites—*contd.*
  radiography of, 1–23
    neutron, 18
    opaque penetrants, 12–16
    positron annihilation, 18–20
    stereo, 16–17
    X-ray, 2–4, 8–9, 20–1
    stress, 20–1
    through-thickness damage, 15–17
Fick's law, 18
Filmless techniques, in radiography, 10
  fluoroscope screen, 10
  xeroradiography, 10
Flaws, fibre, 11–12
Fluorescent screens, 6, 10
  disadvantages of, 6
Fourier analysis, band-selectable. *See*
    ZOOM analysis
Fourier-transform infra-red
    spectroscopy (FTIR), 239–48
  advantage of, 240
  epoxy composite degradation, and,
    242–5
  epoxy resin, and, 240–2
  fibre/matrix interface, and, 245–8
  fibre stress, and, 245
  modes of, 240
Fourier transform NMR, double
    quantum, 210
Fracto-emission (FE), 254–8
Fracture toughness test (K1C), 26, 28,
    30
Free induction decay (FID), 214–15
Frequency
  AE, in, 36, 54
  in-service testing, 180–2
  Larmor, 208
  natural, measuring, 177–82
  quality assurance, 178–80
Fringes
  analysis, automated, 125–9
  Young's, 127, 142
  see also Projected fringes
FRP. *See* Fibre-reinforced plastic
    composites
FTIR. *See* Fourier-transform infra-
    red spectroscopy

γ-aminopropyltriethoxysilane,
    217–18
γ-rays, 1–2, 4–6
  dosage factor, 5
  limits of, 5
  units for measuring, 4–5
  X-rays, and, 4
Geometric fuzziness, 7–8
Glass bottles, and QA, 46
Glass fibre, 11, 20, 214–15
  APS, and, 217–18
  coupling agents, and FTIR, 245–8
  ESR of, 233–5
  NMR, 211–13
  reinforced plastics (GFRP), 25,
    253
    AE, and, 52
  thermography of, 183–5
  vibration, 168, 171, 173, 175
Graham's multiparameter AE, 54–5
Graphite/epoxy laminates, 15–16
Gratings, and moiré systems
  bar-and-space, 107, 119
Grip design, 50

Heat-transfer, mathematics of, 69–
    79
  see also Thermal NDT methods
Heterodyne holography, 135
Holographic correlation, 137
Holography, 108, 127–37, 185–6
  acoustical, 136–7
  classical, 130–6
  heterodyne, 136
  NDT, 133–4
  principles of, 130–2
  problems with, 132–3
Hydrocarbons, 11–12

Image processing, 109, 123–30
  fringe analysis, automated, 125–9
  technology for, 124–5
Image quality indicator (IQI), 8
Impedance method, in NDT, 189–93
Infra-red photography, 66, 68, 87
  Thermovision, 94

Infra-red spectroscopy, Fourier-transform. *See* Fourier-transform infra-red spectroscopy
Injection moulding, 100–1
In-service testing, and natural frequency, 180–2
Interferometry, 35
  holographic. *See* Holography
  moiré. *See* moiré fringe methods
  'speckle', 107, 139–41
Interlaminar shear strength, 52–3
IQI. *See* Image quality indicator
Iridium-192, 5
Isopropyl alcohol, 12

J-integral fracture toughness, 128–9

Kaiser effect, 30–1, 47, 50–1, 57
Kevlar, 25, 53, 214–15, 219, 239
  ESR of, 229–33

Laminated plates, 167–71
  beams cut from, 167–9
Larmor frequency, 208
Lasers, 108, 112–13
  slope measurement, and, 113
LC. *See* Liquid crystals
Lead screens, 6
Liquid crystals, 68, 84–9
  thermogram of, 86
Local vibration measurement, 183–95
  excitation at a single point, 183–6
  excitation at each test point, 186–96
Longitudinal rod mode. *See* Rod mode
Luminescence, 249–54
  chemi-, 250–1
  mechano-, 251–4

Magic-angle spinning (MAS), 210, 217
Martensitic transformation, 26

MAS. *See* Magic-angle spinning
Mechanoluminescence, 251–4
Membrane resonance methods, in NDT, 193–5
ML. *See* Mechanoluminescence
Mode conversion, in AE, 34–5
Moiré fringe methods, 107, 113–23
  bar-and-space gratings, 118
  hardware for, 117
  in-plane deformation, 123
  interferometry, 107, 119, 120–3
  out-of-plane deformation, 113–17
  projection, fringe, 107, 114–17
  shadow, 107, 113–14
  shape deformation, 113–17
Moisture determination, 18–20, 56
  spectroscopy, and, 215–16
Molecular orientation, by thermal NDT, 89–101
Moulding, injection, 100–1

Narmco 5208, 19
Natural frequency. *See* Frequency
NDE (non-destructive evaluation/examination). *See* Non-destructive testing technique (NDT)
NDI (non-destructive inspection). *See* Non-destructive testing technique (NDT)
NDT. *See* Non-destructive testing technique
Neutron radiography, 5, 11, 17
NMR. *See* Nuclear magnetic resonance
Non-destructive testing technique (NDT)
  acoustic emission, 25–63
  chemical spectroscopy, 207–70
  corona discharge, 201–5
  holography, 133–4
  optical methods, 105–49
  radiography, 1–23
  reinforced plastic components, aims, 105–8
  thermal, 65–103

Non-destructive testing technique (NDT)—*contd.*
  vibration methods, 151–200
    acoustic spectra, 195
    amplitude measurement, 185–6
    coin-tap test, 187–9
    impedance, 189–93
    membrane resonance, 193–5
    natural frequency. *See* Frequency
    velocimetry, 195
    vibrothermography, 183–5
NQR. *See* Nuclear quadrupole resonance
Nuclear magnetic resonance, 208–25
  composites, 211–23
  computed tomography, 222–3
  deuterium double quantum decoupling, 211
  double quantum Fourier transform, 210
  MAS, 210, 217–18
  multiple pulse line narrowing, 210
  proton-enhanced (cross polarisation), 209–10, 217
  resonance, nuclear quadrupole, 223–5
Nuclear quadrupole resonance (NQR), 223–5
Numerical modelling, and thermal NDT, 69–79
Nyquist plot, 161–2

Object displacement, in stereoradiography, 16–17
Opaque penetrants, 12–16
Optical fibre crack monitor, 146
Optical methods, for NDT, 35, 105–49
  brittle lacquer, 144–5
  holography, 130–7
  image processing, 123–30
  lasers, 112–13
  moiré fringes, 113–23
  optical fibre crack monitoring, 146
  photoelastic coating, 145–6
  shearography, 143–4
  'speckle', 137
  visual inspection, 108–12

Organic fibre-reinforced plastics (OFRP), 253–4
Orthopositronium. *See* Positron annihilation

Parapositronium. *See* Positron annihilation
Passive heating, in thermal NDT methods, 66–8
  EATF, 67
  heat sources, 67–8
  temperature imaging, 68
Perspex, 80–4, 95
Phase grating, 118, 120
  replication, 118
Photo Flo 600, 12
Photoelastic coating, 145–6
  composites, 146
Photothermal testing. *See* Temperature visualisation
'Pick-off' process, 18–19
PIE. *See* Positive ion emission
Polydiacetylene fibres, 238–9
Polyheteroarylene (PHA), 219, 247
Polyimide resin, 220–1
Polymers, orientation in, 91–101
Polypyromellitimide, 220
Positive ion emission (PIE), 254–5, 257
Positron annihilation, 18–20
Profile projector, 109
Projected fringes, 107
Proof testing, in AE, 47
Propagation, acoustic, 34–5
Proton-enhanced nuclear induction spectroscopy, 209–10

Quality assurance, 59
  AE, 46, 57
  natural frequency, 178–80

Radioactive isotopes, 18
  see also specific names of
Radiography, 1–23
  computer-assisted, 10
  exposure, 8

Radiography—*contd.*
  filmless techniques, 10
  γ-rays, 1–2, 4–5
  geometric fuzziness, 7–8
  neutron, 5, 11, 18
  quality, 6–8
  screens, 6
  thermography, 92
  units, 3–5
  volume fraction, 11
  xero-, 10
  X-ray, 2–4, 8–9, 20–1
Raman spectroscopy, 225–39
  composites, and, 236–9
  resonance (RRS), 238–9
  UV resonance, 237–8
Resonance, UV, 237–8
  see also Membrane resonance, in NDT
Resonant sensors, 36
Ringdown counting, in AE, 40, 42
Rise times, 89
  AE, in, 42
RMS. *See* Root-mean-square method, in AE
Rod mode, 34–5
  conversion, 34–5
  metals, 34
Röntgen rays. *See* Radiography
Root-mean-square method, 45, 58
  AE, 40
RRS. *See* Raman spectroscopy, resonance

Screens, 5–6
  fluorescent, 6
  lead, 6
  neutron radiography, 5
Sensors
  AE, for, 27, 35–6, 38
  broadband, 36
  calibrating, 36
  construction, 36
  electromagnetic interference, 36
  intersensor distances, 38
  resonant, 36
Shadow moiré, 107, 113–14

Shape measurement, three-dimensional, 109–112
  triangulation, 107, 110, 111
Shearography, 107, 143–4
  advantages, 143–4
  disadvantages, 144
Signal/noise, 50
Silicon-29, 211
Slope measurements, and lasers, 113
Sodium-22, 19
Software, computer, 112
  AE recording, 45
  thermal NDT, 79–81, 83–4
Sound, spectrum of, 37
  see also Acoustic emission
Source location, in AE, 42
SP-250, 214
SPATE, 66
Specklography, 126–7, 137–43
  NDT, composites, 143
  principles, 137–9
  techniques, 139–42
    interferometry, 139–41
    photography, 141–2
Spectroscopy. *See* Chemical spectroscopy
Spin-lattice time, 213–14
Static electricity, 10
Stereo radiography, 16–17
  object displacement, 16
Strain
  contours, and moiré, 122–3
  in-plane, and moiré, 117–23
Stress
  fibre, and FTIR, 245
  pattern analysis, by thermal emission. *See* SPATE
  wave factor, 48, 58
  waves, in AE, 26, 28
Strontium-90, 205
Structures
  damping measurements, 176–8
  local vibration, 183–95
  vibration, 160–78
    continuous, 163–71
    single degree of freedom, 160–3
SWF. *See* Stress, wave factor

TBE. *See* Tetrabromoethane
Temperature visualisation, 66, 68
   CFRP, 68
   GRP, 68
   SPATE, 66
Tetrabromoethane, 12–14
Thermal NDT methods, 65–103
   active heating, 65–6
   computer software, 79–81, 83–4
   modelling, 69–84
   molecular orientation, 89–101
   passive heating, 66–8
   perspex, 80–4
Thermal shock, ceramics, 26
Thermographic computer modelling, 69–79
Thermovision-780, 94
'Tin cry'. *See* Twinning
Tomography. *See* Computer-aided tomography
Tread depth, 111–112
Triangulation, 107, 110–12
Twinning, 26, 43
Tyre tread depth, 111–12

Ultrasonics, spectrum, 37
Ultraviolet
   resonance Raman spectroscopy, 237–8
   spectroscopy, 248–9
Unidirectional composites, vibration and, 154–60

Velocimetry, 195
Vibration, structure characteristics, 160–78
   continuous structures, 163–71, 176
   cracks, effects of, 171–5
   damping, 171–5
   dynamic modulus, 171–5
   single degree of freedom, 160–3, 176

Vibration techniques, in NDT, 151–200
   local measurements, 183–95
   acoustic spectra, 195
   amplitude, 185
   coin-tap test, 187–9
   impedance, 189–93
   membrane resonance, 193–5
   velocimetry, 195
   vibrothermography, 183–5
Vibrothermography, 183–5
Vinyl triethoxysilane, 236–7
Vitreosil-xenon photographic flash tube, 87
Volume fractions, 11, 21
VTES. *See* Vinyl triethoxysilane

Watch springs, 46
Water. *See* Moisture determination

Xeroradiography, 10
X-rays, and NDT, 1–5, 10–15, 17, 53
   composite-stress determinations, 20–1
   delamination, 12
   diffraction, 8–9, 20–1
   fibre-content determination, 20–1
   filmless technique, 10
   principles, 2–4
   production, 2–3
   tubes, 2–3

Young's fringes, 127, 142
Young's modulus, 156–60, 163, 165, 168
   AE, 34

Zinc iodide, 12
ZOOM analysis, 177

979/4/1   L1624